Acta Physica Austriaca

Supplementum XVI

Proceedings of the Symposium
"Quantum Dynamics: Models and Mathematics"
at the Centre for Interdisciplinary Research,
Bielefeld University, Federal Republic of Germany,
September 8—12, 1975

1976

Springer-Verlag
Wien New York

Quantum Dynamics: Models and Mathematics

Edited by Ludwig Streit, Bielefeld

With 13 Figures

1976

Springer-Verlag
Wien New York

Prof. Dr. Ludwig Streit
Fakultät für Physik
Universität Bielefeld
Bundesrepublik Deutschland

Softcover reprint of the hardcover 1st edition 1976

Library of Congress Cataloging in Publication Data
Main entry under title:

Quantum dynamics.

(Acta physica Austriaca : Supplementum ; 16)
"Proceedings of the symposium 'Quantum dynamics: models and mathematics,' at the Centre for Interdisciplinary Research, Bielefeld University, Federal Republic of Germany, September 8-12, 1975."
1. Quantum field theory--Congresses. 2. Mathematical physics--Congresses. I. Streit, Ludwig, 1938- II. Bielefeld. Universität. Zentrum für Interdisziplinäre Forschung. III. Series.
QC174.45.A1Q36 536.1'4 76-50623

ISBN-13: 978-3-7091-8475-2 e-ISBN-13: 978-3-7091-8473-8
DOI: 10.1007/978-3-7091-8473-8

Preface

The contributions to this volume deal with topics ranging over constructive and general quantum field theory and related algebraic problems, non-renormalizable models, equilibrium statistical mechanics, critical phenomena, and nonlinear equations modelling the onset of turbulence.

They are based on lectures intended to provide the 1975/1976 research group "Mathematical Problems of Quantum Dynamics" at the Centre for Interdisciplinary Research (ZiF) of Bielefeld University with an input reflecting important recent developments and presented by leading experts in the pertinent fields of research.

They further reflect a situation of unusually active and fruitful exchange not only between various specializations of theoretical physics which deal with the specific problems of large systems but also of a lively two-way interaction with mathematics which stimulates and furthers the progress of both disciplines.

Thanks are due to the contributors, to the Preparatory Committee - H. Behncke, P. Blanchard, K. Hepp, O. Steinmann, A.S.Wightman -, to the University of Bielefeld for the sponsorship of these lectures, to the directors and staff of ZiF who made them possible, and to Miss V.C. Fulland and Miss M. Kämper for their calm and competent production of the manuscript.

Bielefeld, October 1976 Ludwig Streit

Contents

The Lecturers

H.J. Borchers
Universität Göttingen

J. Dimock
Université de Genève

Jean-Pierre Eckmann
Université de Genève

James Glimm
Rockefeller University, New York

Arthur Jaffe
Harvard University, Cambridge, Mass.

Francesco Guerra
Universita di Salerno

John R. Klauder
Bell Laboratories, Murray Hill, N.J.

Joel L. Lebowitz
Yeshiva University, New York

Oliver A. McBryan
Rockefeller University, New York

Konrad Osterwalder
Harvard University, Cambridge, Mass.

David Ruelle
IHES, Bures-sur-Yvette

R. Sénéor
Ecole Polytechnique, Palaiseau

Thomas Spencer
Rockefeller University, New York

Erling Størmer
University of Oslo

K. Symanzik
DESY, Hamburg

Acta Physica Austriaca, Suppl. XVI, 1–14 (1976)

Jordan Algebras Versus C^*- Algebras

Erling Størmer
Department of Mathematics,
University of Oslo

1. Introduction

In the more abstract setting for quantum statistical mechanics and quantum field theory C^*-algebras have become quite popular. They are useful for explaining several phenomena, and the mathematics is well developed, so the popularity is very natural. However, if one looks at them from a more axiomatic point of view their use is not so clear. For example, in quantum mechanics there has been and still is disagreement on the formulation of the basic axioms, and especially whether one can right away assume the observables form self-adjoint operators on a Hilbert space.

In the present notes we shall study this problem taking an axiomatic position somewhere in the middle, and then see how close we are to considering C^* - algebras. We shall assume the observables form an abstract Jordan algebra with sufficiently extra structure so that we can do spectral theory. Then we shall discuss a recent theorem of Alfsen, Shultz and myself [2], which tells us that the Jordan algebras in question are so close to being the self-adjoint parts of C^*-algebras, that the study of the latter is quite well founded.

2. Basics

In a paper almost 30 years ago Segal [6] proposed axioms close to those I indicated in the introduction. In addition to assumptions on a norm he assumed the observables formed a real vector space in which he could form polynomials, and in particular squares. With such assumptions one can define the Jordan product between two observables a and b:

$$a \circ b = \frac{1}{2}\left((a+b)^2 - a^2 - b^2\right). \qquad (1)$$

With his assumptions Segal could do spectral theory, but as pointed out by Sherman [7] one needed stronger assumptions on the product (1) in order to exclude uninteresting cases. We shall therefore assume the observables form a real vector space in which squares are defined so that the product (1) is distributive and satisfies the weak associativity

$$a^2 \circ (b \circ a) = (a^2 \circ b) \circ a,$$

i.e. the observables form a Jordan algebra (over the reals $\mathbb{R}$).

More than 40 years ago Jordan, von Neumann, and Wigner [5] classified the finite dimensional simple Jordan algebras with the additional property

$$a^2 + b^2 + \ldots + c^2 = 0 \Rightarrow a = b = \ldots = c = 0. \qquad (2)$$

They showed that the examples are

(i) the hermitian $n \times n$ matrices over the reals, complexes, and quaternions,

(ii) the hermitian 3×3 matrices over the Cayley numbers, denoted by M_3^8, and often called the exceptional algebra.

(iii) the finite dimensional spin factors, in which the identity is never the sum of more than two idempotents. These Jordan algebras will be discussed in more detail below.

It was shown by Albert and Paige [1] that M_3^8 has no Jordan representations on a real or complex or quaternionic Hilbert space, hence we already have an axiomatically natural Jordan algebra which is not representable on Hilbert space. This exceptional algebra is what makes representation theory for Jordan algebras difficult.

From the point of view of representation theory the nice Banach *-algebras are the B^*- algebras, namely those Banach *- algebras over the complex numbers in which the norm satisfies $||x^*x|| = ||x||^2$ for all x. The famous Gelfand-Neumark theorem asserts that a B^*- algebra has a faithful representation as a C^*- algebra of operators on a complex Hilbert space. The Jordan analogue of B^*- algebras is given in

Definition 1.
A JB- algebra is a Jordan algebra (over $\mathbb{R}$) with identity 1, which is a Banach space with respect to a norm satisfying for all a,b,

(i) $||a \circ b|| \leq ||a|| \, ||b||$,

(ii) $||a^2|| = ||a||^2$,

(iii) $||a^2|| \leq ||a^2+b^2||$.

I'll take the view that the observables form a JB- algebra. The axioms (i)-(iii) on the norm are polished versions of those introduced by Segal [6]. Axiom (iii) may look a little ad hoc. It is necessary because we want at least the reality condition (2) to hold, and because the B^*- condition $||x^*x|| = ||x||^2$ is much stronger that (ii). If we let A be the set of functions f analytic for $|z| < 1$, continuous for $|z| = 1$, and real for z real, then with pointwise multiplication and norm given by

$||f|| = \sup_{|z|\leq 1} |f(z)|$, A is a Jordan algebra satisfying (i), (ii), and (2), but not (iii). Since we want to exclude A from consideration we must assume (iii).

Two examples of JB- algebras are well known, namely M_3^8, see [7], and norm closed Jordan algebras of self-adjoint operators on complex Hilbert space. The latter algebras were called JC-algebras by Topping. By (1) the Jordan product is $a \circ b = \frac{1}{2}(ab+ba)$. JC-algebras were studied by Topping and myself about 10 years ago, see [8] - [12], and also [3]. The irreducible ones were classified as follows.

a) A is the self-adjoint part of a C^*- algebra.

b) A has a faithful representation as a JB- algebra consisting of hermitian operators on a real or quaternionic Hilbert space. In this case let R(A) be the norm closed real *- algebra generated by A, and let $iR(A) = \{ia : a \in R(A)\}$. Then

$A = R(A)_{s.a.}$ is the self-adjoint part of R(A),

$R(A) \cap iR(A) = \{0\}$,

$R(A) + iR(A) = \{a+ib : a,b \in R(A)\}$, is a C^*- algebra.

Thus A is not far from being the self-adjoint part of a C^*-algebra.

c) A is a spin factor. The abstract definition is as follows. Let H be a real Hilbert space of dimension ≥ 3, and let e be a distinguished unit vector in H. Let $N = \{e\}^\perp$, so $H = \mathbb{R}e \oplus N$. Then H is by definition a spin factor when the product is defined by

$$(\alpha e \oplus a) \circ (\beta e \oplus b) = (\alpha\beta + (a,b))\, e \oplus (\beta a + \alpha b),$$

$$\alpha,\beta \in \mathbb{R}, \quad a,b \in N.$$

The prototype is well-known and arises from the anti-commutation relations. Let H be a real Hilbert space and $f \to a(f)$ a representation of H satisfying the canonical anti-commutation relations, i.e.

$$a(f)a(g)^* + a(g)^*a(f) = (g,f)1$$

$$a(f)a(g) + a(g)a(f) = 0.$$

Let

$$b(f) = a(f) + a(f)^*,$$

$$c(f) = i(a(f)-a(f)^*).$$

Then b(f) and c(f) are self-adjoint elements of the CAR- algebra. Computing Jordan products we have $b(f) \circ b(g) = (f,g)1$, $c(f) \circ c(g) = (f,g)1$, and $b(f) \circ c(g) = 0$, for all $f,g \in H$. Let A be the norm closed linear span of 1, b(f), c(g), $f,g \in H$. Then A is a JC- algebra generating the CAR- algebra. Moreover, if we define an inner product on A by

$$(\alpha 1 + \sum_r b(f_r) + \sum_s c(g_s), \beta 1 + \sum_t b(h_t) + \sum_u c(k_u)) =$$

$$= \alpha\beta + \sum_{r,t} (f_r, h_t) + \sum_{s,u} (g_s, k_u),$$

it is clear that with e = 1 and N the norm closed linear span of b(f), c(g), f, $g \in H$, $A = \mathbb{R} e + N$ is a spin factor, and from counting dimensions it is clear that all spin factors are obtained in this way.

Recently Alfsen, Shultz and I have shown that all JB-algebras are combinations of the above examples [2]. To formulate the theorem, a <u>Jordan ideal</u> in a JB- algebra A is a norm closed linear subspace J of A such that $a \in A$, $b \in J \Rightarrow a \circ b \in J$. It can be shown just as for C^*- algebras that the quotient A/J with its natural algebraic structure and the quotient norm, is a JB- algebra.

<u>Theorem</u>.
Let A be a JB- algebra. Then there exists a unique Jordan ideal J in A such that A/J has a faithful Jordan representation as a JC- algebra, while every "irreducible" representation of A not annihilating J is onto the exceptional algebra M_3^8.

The term "irreducible" will be made more precise in the next section.

3. The Proof of the Theorem

In this section I'll outline the proof of the theorem. The details are somewhat lengthy, and can be found in [2]. The first step in the proof is to give an order theoretic definition of JB- algebras. Recall that an order unit space (A,1) is an ordered vector space over the reals with an order unit 1, which is archimedean in that $a \leq \frac{1}{n}1$, $n = 1,2,\ldots$, implies $a \leq 0$, and which has a norm defined by

$$||a|| = \inf \{\lambda>0 : -\lambda 1 \leq a \leq \lambda 1\}.$$

Proposition 1

A normed Jordan algebra A over the reals with identity 1 is a JB-algebra if and only if (A,1) is a complete order unit space such that $-1 \leq a \leq 1 \Rightarrow 0 \leq a^2 \leq 1$.

From this result it follows that the positive cone A^+ in a JB-algebra A is the set of squares. Furthermore, standard techniques show that we can do spectral theory and that A has enough states.

Since the non-assosiativity of a JB- algebra makes it impossible to obtain a GNS- construction for its states, our proof will not follow the pattern of the Gelfand-Neumark theorem. To obtain more structure we shall imbed a JB- algebra A in its second dual A^{**}, and hope that pure states of A correspond to minimal central projections in A^{**}, so that the representation $A \to A^{**}e$, e minimal central projection, is well defined, and $A^{**}e$ is so nice that it can be classified. Then we want to take the direct sum of these representations. Since we have been unable to show that A^{**} is a JB- algebra we must be more careful.

Let A be a JB- algebra considered as a subset of its second dual A^{**}. If ρ is a state of A, defined by Proposition 1, $a \to \langle a^2,\rho\rangle^{1/2}$

is a semi-norm on A. These semi-norms define the strong topology on A. The weak topology on A^{**} is the $\sigma(A^{**},A^{*})$-topology and is defined by the semi-norms $a \to |\langle a,\rho\rangle|$, $\rho \in A^{*}$. We let

$\tilde{A} = \{a \in A^{**}$: a is a weak limit of a bounded strong Cauchy net in A$\}$.

It can be shown that $\tilde{A}$ is a monotone closed JB- algebra such that the states of A extend to normal states on $\tilde{A}$. Furthermore, the lattice structure for its idempotents is just like the projection lattice in a von Neumann algebra. Similarly spectral theory is like that in a von Neumann algebra.

Since multiplication is commutative in a Jordan algebra it is not immediately clear what the center should be. If A is a JB-algebra and $a \in A$ we let L_a denote the operator on A defined by $L_a b = a \circ b$. We say as in [4] that a and b operator commute if $L_a L_b = L_b L_a$. If M is a JB- algebra with the same properties as $\tilde{A}$ we define the center of M to be the set of $a \in M$ such that a operator commutes with all idempotents in M. Then the center is a strongly closed associative subalgebra. We say M is a JB- factor if its center is the scalars. A more easily understood definition of the center can be obtained as follows.

Let A be a Jordan algebra. If $a,b \in A$ we generalize the product a b a in the associative case by defining

$$\{a\ b\ a\} = 2a \circ (a \circ b) - a^2 \circ b.$$

If we denote by U_a the operator $U_a : A \to A$ by $U_a b = \{a\ b\ a\}$, it follows that if A is a JB- algebra then $U_a : A^+ \to A^+$. If p is an idempotent in A then $U_p(A)$ is a JB- algebra, which we shall denote by A_p. However, before we forget the center, we should state

Proposition 2

With M a monotone closed JB- algebra as above, an element a in M belongs to the center of M if and only if $U_s a = a$ for each symmetry $s \in M$, i.e. for each $s \in M$ such that $s^2 = 1$.

This proposition is easy from the definition, since every symmetry s is of the form 2p - 1 with p an idempotent.

We can now obtain the analogue of the representation $A \to A^{**}e$ with e a minimal central idempotent, indicated above. If ρ is a state of A let $c(\rho)$ be the support of the extension of ρ to $\tilde{A}$, restricted to the center of $\tilde{A}$. Thus $c(\rho)$ is the minimal central idempotent p in $\tilde{A}$ such that $\rho(p) = 1$.

Proposition 3
Let A be a JB- algebra and ρ a state of A. Let $\phi_\rho : A \to \tilde{A}_{c(\rho)}$ by $\phi_\rho(a) = U_{c(\rho)}\tilde{a}$, where $\tilde{a}$ is the image of a in $\tilde{A}$. Then ϕ_ρ is a Jordan homorphism with image weakly dense in $\tilde{A}_{c(\rho)}$. Furthermore, if ρ is a pure state then $\tilde{A}_{c(\rho)}$ is a JB- factor.

We have thus arrived at the stage where we have to in some sense or other to classify JB- factors. Except for those which are spin factors, we shall write them as matrix algebras in a way analogous to that von Neumann algebras can be written as matrix algebras. For this purpose we need comparison theory for idempotents in a JB- factor M. If p and q are idempotents in M we say they are equivalent, written $p \sim q$, if there exists a finite family $s_1, \ldots, s_n$ of symmetries in M such that

$$U_{s_n} U_{s_{n-1}} \cdots U_{s_1} p = q.$$

If $q = U_s p$ for a symmetry s we say $p \sim q$ via a symmetry. While equivalence is an equivalence relation, equivalence via a symmetry need not be. However, to get matrix algebras we need equivalence of two projections via a symmetry. The crucial lemma for this purpose was communicated to us by R. Schafer.

Lemma 1
Let p and q be idempotents in a JB- factor M such that $p \circ q = 0$, i.e. p and q are orthogonal. Suppose there exists two symmetries s and t in M such that $U_s U_t p = q$. Then $p \sim q$ via a symmetry.

From this lemma it is straightforward to obtain the important "halving lemma".

Proposition 4

Let M be a JB- factor without minimal idempotents. Then every idempotent e in M can be halved, i.e. there exist idempotents p and q in M with $e = p+q$ and $p \sim q$ via a symmetry.

The formal definition of our matrix algebras is as follows. Let $\mathcal{a}$ be an algebra over the reals with identity 1 and involution *. Let $\mathcal{a}_n$ be the n x n matrices over $\mathcal{a}$ with involution $x \to x^*$ defined by $(a_{ij})^* = (a_{ji}^*)$, and let $H(\mathcal{a}_n)$ be the hermitian matrices in $\mathcal{a}_n$ with Jordan product $x \circ y = \frac{1}{2}(xy+yx)$. If $H(\mathcal{a}_n)$ is a Jordan algebra it is called a Jordan matrix algebra.

If n is a positive integer we say a JB- factor is of type I_n if the identity is the sum of n (orthogonal) minimal idempotents. It is not hard to show that those of type I_2 are spin factors.

Proposition 5

Every JB- factor M except those of type I_2 is isomorphic to a Jordan matrix algebra $H(\mathcal{a}_n)$. If in addition M is not of type I_3 then $\mathcal{a}$ is associative.

To prove the proposition if there are no minimal idempotents in M we use the halving lemma (Prop.4) twice to show that the identity is the sum of 4 orthogonal idempotents each pair of which are equivalent via a symmetry. Then a theorem of Jacobson [4] shows that M is of the form $H(\mathcal{a}_4)$. The I_n case gives us a Jordan matrix algebra $H(\mathcal{a}_n)$, and if the identity is the sum of an infinite number of orthogonal idempotents we show M is of the form $H(\mathcal{a}_4)$. The final statement follows from another theorem of Jacobson, see [4].

The classical results of Jordan, von Neumann, and Wigner [5] now enable us to restrict attention to the case $M = H(\mathcal{a}_n)$ with $\mathcal{a}$ associative. Then we can define a norm on $\mathcal{a}_n$ by $||x|| = ||x^*x||^{1/2}$, so $\mathcal{a}_n$ becomes a real Banach algebra with a B^*- condition on its norm. If now ρ is a state on M it can be extended to be a state on $\mathcal{a}_n$ by letting it be zero on skew-adjoint elements. Thus the GNS- construction can be performed, and we get a representation of M on a real Hilbert space, and thus easily on a complex Hilbert space. We thus have

<u>Proposition 6</u>
Every JB- factor except M_3^8 is isomorphic to JC- algebra.

M_3^8 can be distinguished algebraically from JC- algebras by the so-called s- <u>identities</u>, see [4]. These are identities in three variables of degree at least 8, which hold in JC- algebras but not in M_3^8. Let $f(a,b,c) = 0$ be one such identity. From the last proposition it is easy to obtain the following Gelfand-Neumark theorem for JB- algebras.

<u>Proposition 7</u>
A JB- algebra A has a faithful representation as a JC- algebra if and only if $f(a,b,c) = 0$ for all $a,b,c \in A$.

In order to conclude the proof of the main theorem we let J be the Jordan ideal generated by all elements of the form $f(a,b,c)$ with $a,b,c \in A$. Then in the quotient A/J we have $f(a,b,c) = 0$, so by Proposition 7, A/J has a faithful representation as a JC-algebra. Conversely, if ϕ is a representation of A into $\tilde{A}$ such that the strong closure of $\phi(A)$ is a JB- factor and $\phi(J) \neq 0$, then from Proposition 6 and 7 $\phi(A) = M_3^8$. This last remark also explains the word "irreducible" in the statement of the theorem.

4. Conclusions

Coming back to the discussion in the introduction to this article I would like to draw the follwing conclusions from the results of the previous two sections.

If we assume the observables form a Jordan algebra with enough analytic structure so we can do spectral theory, we may as well assume they form a JB- algebra. Since in a JB- algebra the exceptional algebra M_3^8 is contained in an ideal I feel that natural physical assumptions like locality would exclude M_3^8 from appearing in the JB- algebra. However, algebraic assumptions should not exclude M_3^8, since s - identities are algebraically unnatural. Thus, in physical problems I would find it natural to assume the observables form a JC- algebra. From the examples we gave for JC- algebras there are essentially two cases. The study of spin factors was the same as that of the canonical anticommutation relations, hence of the well known CAR- algebra. The other JC-algebras are very close to being the self-adjoint parts of C^*-algebras. Thus I do not think any information is lost if we assume the JB- algebra of observables is the self-adjoint part of a C^*- algebra.

References

1) A.A. ALBERT and L.J. PAIGE, On a Homomorphism Property of Certain Jordan Algebras, Trans. Amer. Math. Soc. 93 (1959), 20-29.
2) E. ALFSEN, F. SHULTZ, and E. STØRMER, A Gelfand-Neumark Theorem for Jordan Algebras, (to appear in Advances in Math.)
3) E. EFFROS and E. STØRMER, Jordan Algebras of Self-Adjoint Operators, Trans. Amer. Math. Soc. 127 (1967), 313-316.
4) N. JACOBSON, Structure and Representations of Jordan Algebras, Amer. Math. Soc. Colloq. Publ. 39. Amer. Math. Soc. Providence 1968,
5) P. JORDAN, J. VON NEUMANN, and E. WIGNER, On an Algebraic Generalization of the Quantum Mechanical Formalism, Ann. Math. 35 (1934), 29-64.
6) I.E. SEGAL, Postulates for General Quantum Mechanics, Ann. Math. 48 (1947), 930-948.
7) S. SHERMAN, On Segal's Postulates for General Quantum Mechanics, Ann.Math. 64 (1956), 593-601.
8) E. STØRMER, On the Jordan Structure of C^*- Algebras, Trans. Amer. Math. Soc. 120 (1965), 438-447.
9) ———, Jordan Algebras of Type I, Acta Math. 115 (1966), 165-184.
10) ———, Irreducible Jordan Algebras of Self-Adjoint Operators, Trans. Amer. Math. Soc. 130 (1968), 153-166.
11) D. TOPPING, Jordan Algebras of Self-Adjoint Operators, Mem. Amer. Math. Soc. 53 (1965).
12) ———, An Isomorphism Invariant for Spin Factors, J.Math. Mech. 15 (1966), 1055-1064.

Acta Physica Austriaca, Suppl. XVI, 15–46 (1976)

Decomposition of Families of Unbounded Operators

H.J. Borchers
Institut für Theoretische Physik
Universität Göttingen

Contents:

I. Introduction

In this seminar I will talk about a joint work with J. Yngvason, which has been published in the Communications in Mathematical Physics in two papers with the titles:

> On the Algebra of Field Operators. The Weak Commutant and Integral Decomposition of States. C.M.P. 42, 231 (1975)

and:

> Integral Representations for Schwinger Functionals and the Moment Problem over Nuclear Spaces. C.M.P. 43, 255 (1975)

Our investigation was originated by the following question: Let A be a * - algebra with identity and let ω be a state on A this means ω is a normalized positive linear functional on A. When can such a functional be decomposed into extremal ones? This leads first to the problem of characterizing extremal states. For the case of one single symmetric operator, it is known that it is not at all necessary to investigate the algebra generated by this operator, but, one gets along by looking at this operator alone.

This leads us to the

I.1. Definition: A partial * - algebra is a complex vector space A together with an involution $x \in A \to x^* \in A$ with the usual operation $(x+\lambda y)^* = x^* + \bar{\lambda} y^*$, $x^{**} = x$. And a subset $M \subset A \times A$ such that $(x,y) \in M$ implies $(y^*,x^*) \in M$, (x,y_1) and $(x,y_2) \in M$ implies $(x,y_1+y_2) \in M$ and for every pair $(x,y) \in M$ exist an element $x \cdot y \in A$ fulfilling the usual operations.

I.2. Definition: Let A be a partial * - algebra, a pair $(\pi, \mathcal{D})$ is called a * - representation of A if the following conditions are fulfilled:

(i) $\mathcal{D}$ is a pre-Hilbert-space with completion $\mathcal{H}(\mathcal{D})$

(ii) To every $x \in A$ exists alinear operator $\pi(x)$ defined on $\mathcal{D}$ with values in $\mathcal{H}(\mathcal{D})$ such that

(α) $\pi(x+\lambda y) = \pi(x) + \lambda\pi(y)$

(β) if $(x,y) \in M$ then we have $\pi(y)\mathcal{D} \subset \mathcal{D}$ and $\pi(xy) = \pi(x)\pi(y)$

(γ) for $f,g \in \mathcal{D}$ we have the relation $(f,\pi(x)g) = (\pi(x^*)f,g)$.

<u>Remarks</u>

1) The last condition implies, that every operator $\pi(x)$ is closable.
2) It is not assumed that the common domain $\mathcal{D}$ is closed in any topology, also not in the topology induced by the graph norms of all the operators $\pi(x)$.
3) If A is a * - algebra and ω a state on A, then one gets $(\pi, \mathcal{D})$ in the usual way by the G.N.S. construction.

If we want to define extremal states, then we have to speak about the commutant of the representation. But, in the case of unbounded operators we have to distinguish between two different kinds of commutants.

I.3. <u>Definition:</u> Let $(\pi, \mathcal{D})$ be a * - representation of a partial * - algebra A then we define

(a) The strong commutant:

$$(\pi, \mathcal{D})'_s = \{c \in B(\mathcal{H}(\mathcal{D})); c\mathcal{D} \subset \mathcal{D} \text{ and } \pi(x)c = c\pi(x) \text{ on } \mathcal{D}$$
$$\text{for all } x \in A\}$$

(b) The weak commutant:

$$(\pi, \mathcal{D})'_w = \{c \in B(\mathcal{H}(\mathcal{D})); (cf,\pi(x)g) = (\pi(x^*)f,c^*g)$$
$$\text{for all } f,g \in \mathcal{D} \text{ and all } x \in A\}$$

With this notation we can characterize extremal states, namely a state on a * - algebra A is extremal if and only if the weak commutant of the cyclic representation π_ω is trivial i.e. consists of scalar multiples of the identity. This statement follows trivially from the properties of the weak commutant which are listed in the following

I.4. <u>Lemma:</u> Let $(\pi, \mathcal{D})$ be a $*$ - representation of a partial $*$ - algebra then

(a) The strong commutant $(\pi, \mathcal{D})'_s$ is an algebra

(b) The weak commutant $(\pi, \mathcal{D})'_w$ is

(i) weakly closed and contains the identity

(ii) invariant under the adjoint operation, i.e. $c \in (\pi, \mathcal{D})'_w$ implies $c^* \in (\pi, \mathcal{D})'_w$

(iii) is generated by its positive elements

(iv) $(\pi, \mathcal{D})'_s \subset (\pi, \mathcal{D})'_w$.

All these properties follow easily from the definitions, so that we will not give the proofs. But we will make some

<u>Remarks</u>

1) $(\pi, \mathcal{D})'_s$ is an algebra, but in the general stituation it is not a $*$ - algebra and is also not closed in any reasonable operator topology.

2) $(\pi, \mathcal{D})'_w$ is not an algebra in general.
If you call $\bar{\mathcal{D}} = \bigcap_{x \in A} \mathcal{D}_{\overline{\pi(x)}}$ then $(\pi, \mathcal{D})$ has an extension $(\bar{\pi}, \bar{\mathcal{D}})$ by continuity. It is simple to show that $(\pi, \mathcal{D})$ and $(\bar{\pi}, \bar{\mathcal{D}})$ have the same weak commutant.

3) If one deals with a family of bounded operators, then one can put $\mathcal{D} = \mathcal{H}(\mathcal{D})$. In this case, the strong and weak commutants coincide.

If you now look at the decomposition theory for bounded operators then it consists of two things

Step 1. In this case you have only one kind of commutant which is a von Neumann algebra. Pick a maximal abelian algebra M in this commutant.

Step 2. Try to define an integral decomposition with respect to M, which is always possible if the Hilbert space is separable.

This leads for unbounded operators to the following two questions

1) How can you find a maximal commuting algebra, since one wants to make a decomposition with respect to the weak commutant?
2) Assume you can construct a maximal commuting algebra in the weak commutant which additional information is needed in order to define an integral decomposition?

Both questions will be treated separately.

II. Extension Theory

As a guide let us consider the case of one symmetric operator. In this case one would try to find a self-adjoint extension in order to find a maximal abelian subalgebra in the commutant. But in the case that this operator has non symmetric defect indices, we can define selfadjoint extensions only in some enlarged Hilbert space. So the first step will be to look for extensions of a family of unbounded operators.

II.1. Definition: Let A be a partial * - algebra and $(\pi, \mathcal{D})$ a representation of A. A representation $(\hat{\pi}, \hat{\mathcal{D}})$ will be called an extension of $(\pi, \mathcal{D})$ if:

(i) $(\hat{\pi}, \hat{\mathcal{D}})$ is a * - representation of A

(ii) $\hat{\mathcal{D}} \supset \mathcal{D}$ and the norm on $\hat{\mathcal{D}}$ coincides on $\mathcal{D}$ with the original norm on

(iii) For every $x \in A$ the restriction $\hat{\pi}(x)$ $\mathcal{D}$ is equal to $\pi(x)$.

Remark:

It is not required in this definition that the two Hilbert spaces coincide but from $\mathcal{D} \subset \hat{\mathcal{D}}$ follows that $\mathcal{H}(\mathcal{D})$ is a subspace of $\mathcal{H}(\hat{\mathcal{D}})$.

If $(\hat{\pi}, \hat{\mathcal{D}})$ is an extension of $(\pi, \mathcal{D})$ and b is a bounded linear operator on $\mathcal{H}(\hat{\mathcal{D}})$ then we can define an operator on $\mathcal{H}(\mathcal{D})$ by E b E, where E denotes the projection onto $\mathcal{H}(\mathcal{D})$. Since this mapping occurs quite often in the following, we will introduce a separate notation for it.

II.2. Definition: Let $(\hat{\pi}, \hat{\mathcal{D}})$ be an extension of $(\pi, \mathcal{D})$ and E be the projection onto $\mathcal{H}(\mathcal{D})$ then we define for any bounded operator b on $\mathcal{H}(\hat{\mathcal{D}})$

$$\rho(b) = E\, b\, E$$

II.3. Lemma:

(i) ρ is linear, commutes with the involution and preserves ordering i.e.

$$\rho(x+\lambda y) = \rho(x) + \lambda\rho(y); \ \rho(x^*) = \rho(x)^*$$

and $x \geq y$ implies $\rho(x) \geq \rho(y)$.

(ii) ρ is weakly continuous

(iii) If $b\varepsilon(\hat{\pi}, \hat{\mathcal{D}})'_w$ then follows $\rho(b)\varepsilon(\pi, \mathcal{D})'_w$

These properties are all easy to verify and can be done by the reader.

Remark:

ρ maps the whole weak commutant of $(\hat{\pi}, \hat{\mathcal{D}})$ into the weak commutant of $(\pi, \mathcal{D})$, in particular it maps also the strong commutant of $(\hat{\pi}, \hat{\mathcal{D}})$ into the weak commutant of $(\pi, \mathcal{D})$. This means that if b commutes strongly with $(\hat{\pi}, \hat{\mathcal{D}})$ then $\rho(b)$ need not commute strongly any more with $(\pi, \mathcal{D})$. Therefore the plan of attack consists in the inverse procedure, namely to take elements b from the weak commutant and try to define an extension $(\hat{\pi}, \hat{\mathcal{D}})$ in such a way that $b = \rho(b')$ for some strongly commuting element $b'\varepsilon(\hat{\pi}, \hat{\mathcal{D}})'_s$. This plan leads naturally to the following

II.4. Definition:

1) A triple $(\hat{\pi}, \hat{\mathcal{M}}, \hat{\mathcal{D}})$ will be called an induced extension of $(\pi, \mathcal{D})$ if
   (i) $(\hat{\pi}, \hat{\mathcal{D}})$ is an extension of $(\pi, \mathcal{D})$.
   (ii) $\hat{\mathcal{M}}$ is an abelian $*$ - algebra of bounded operators on $\mathcal{H}(\hat{\mathcal{D}})$ with $1\varepsilon\hat{\mathcal{M}}$ and $\hat{\mathcal{M}} \subset (\hat{\pi}, \hat{\mathcal{D}})'_s$
   (iii) $\hat{\mathcal{D}}$ is the linear span of $\hat{\mathcal{M}}\,\mathcal{D}$ i.e.

$$\hat{\mathcal{D}} = \{\sum m_i f_i;\ m_i \ \varepsilon\, \hat{\mathcal{M}}, f_i \ \varepsilon\, \mathcal{D}\}$$

2) Denote by $\hat{\pi}(A)\vee\hat{\mathcal{M}}$ the linear span of

$$\{\hat{\pi}(x), x\varepsilon A\} \cup \mathcal{M} \cup \{\hat{\pi}(x)m;\ x\varepsilon A,\ m\varepsilon\mathcal{M}\}$$

and assume $(\hat{\pi}, \hat{\mathcal{M}}, \hat{\mathcal{D}})$ and $(\hat{\hat{\pi}}, \hat{\mathcal{N}}, \hat{\hat{\mathcal{D}}})$ are two induced extensions then we introduce a semiordering by

$$(\hat{\pi}, \hat{\mathcal{M}}, \hat{\mathcal{D}}) <\cdot (\hat{\hat{\pi}}, \hat{\mathcal{N}}, \hat{\hat{\mathcal{D}}})$$

if

(i) $\hat{\mathcal{D}} \subset \hat{\hat{\mathcal{D}}}$

(ii) There exists a sub-algebra $\hat{\hat{\mathcal{M}}} \subset \hat{\hat{\mathcal{N}}}$ isomorphic to $\hat{\mathcal{M}}$ such that

(iii) $(\hat{\hat{\pi}} \vee \hat{\hat{\mathcal{M}}}, \hat{\hat{\mathcal{N}}}, \hat{\hat{\mathcal{D}}})$ is an induced extension of $(\hat{\pi} \vee \hat{\mathcal{M}}, \hat{\mathcal{D}})$.

The last definition points out that we want to construct induced extensions several times. Therefore we have to show that this concept is consistent.

II.5. <u>Lemma:</u>

1) Let $(\hat{\pi}, \hat{\mathcal{M}}, \hat{\mathcal{D}})$ be an induced extension of $(\pi, \mathcal{D})$ and assume there exist $x \varepsilon A$ such that $\pi(x)$ is bounded, then follows $||\pi(x)||=||\hat{\pi}(x)||$.

2) Let $(\hat{\pi}, \hat{\mathcal{M}}, \hat{\mathcal{D}})$ be an induced extension of $(\pi, \mathcal{D})$ and denote by $\hat{\mathcal{M}}^-$ the weak closure of $\hat{\mathcal{M}}$ and $\hat{\mathcal{D}}^-$ the linear span of $\hat{\mathcal{M}}^- \mathcal{D}$, then every operator $\hat{\pi}(x)$ has a continuous extension $\hat{\pi}^-(x)$ to and $(\hat{\pi}^-, \hat{\mathcal{M}}^-, \hat{\mathcal{D}}^-) > (\hat{\pi}, \hat{\mathcal{M}}, \hat{\mathcal{D}})$.

Proof: Assume $B \geq 0$ then follows

$$\sum_{ij} (m_i f_i, \hat{B} m_j f_j) = \sum_{ij} (f_i, m_i^* m_j B f_j) =$$

$$=\sum_{ij} (B^{1/2} f_i, m_i^* m_j B^{1/2} f_j)$$

$$=||\sum_i m_i B^{1/2} f_i||^2 \geq 0$$

This shows $\hat{B}$ is positive. If now $\hat{B}$ is selfadjoint then follows the lower and upper bounds of $\hat{B}$ are the same as those of B. Hence $||\hat{B}|| = ||B||$. But this implies that 1) holds for arbitrary elements.

The second statement follows from the usual continuity since the elements $(mf, m\pi(x)f)$; $m \varepsilon \hat{\mathcal{M}}$, $f \varepsilon \mathcal{D}$ belong to the graph of $\hat{\pi}(x)$.

Having an induced extension $(\hat{\pi}, \hat{\mathcal{M}}, \hat{\mathcal{D}})$ then ρ maps $\hat{\mathcal{M}}$ into the weak commutant of $(\pi, \mathcal{D})$. But we can hope to reconstruct $(\hat{\pi}, \hat{\mathcal{M}}, \hat{\mathcal{D}})$ only if ρ is a bijective mapping. Knowing $(\hat{\pi}, \hat{\mathcal{M}}, \hat{\mathcal{D}})$ then we eventually want to construct an extension of this representation $(\hat{\hat{\pi}}, \hat{\mathcal{N}}, \hat{\hat{\mathcal{D}}})$. This can be done hopefully by using operators which weakly commute with $\hat{\pi}$ and $\hat{\mathcal{M}}$. Therefore ρ should also be unique on such elements. This leads us to the

II.6. <u>Definition:</u> Let $(\hat{\pi}, \hat{\mathcal{M}}, \hat{\mathcal{D}})$ be an induced extension of $(\pi, \mathcal{D})$ then

1) we define: $(\hat{\pi}, \hat{\mathcal{M}}, \hat{\mathcal{D}})'_w = (\hat{\pi}, \hat{\mathcal{D}})'_w \cap \hat{\mathcal{M}}'$
2) we say $(\hat{\pi}, \hat{\mathcal{M}}, \hat{\mathcal{D}})$ is regular if the mapping ρ restricted to $(\hat{\pi}, \hat{\mathcal{M}}, \hat{\mathcal{D}})'_w$ is injective.

II.7. <u>Lemma:</u>

1) Let $(\hat{\pi}, \hat{\mathcal{M}}, \hat{\mathcal{D}})$ be a regular induced extension of $(\pi, \mathcal{D})$ and $(\hat{\hat{\pi}} \vee \hat{\mathcal{M}}, \hat{\hat{\mathcal{L}}}, \hat{\hat{\mathcal{D}}})$ be a regular induced extension of $(\hat{\pi} \vee \hat{\mathcal{M}}, \hat{\mathcal{D}})$ then $(\hat{\hat{\pi}}, \hat{\hat{\mathcal{M}}} \vee \hat{\hat{\mathcal{L}}}, \hat{\hat{\mathcal{D}}})$ is a regular induced extension of $(\pi, \mathcal{D})$.
2) Every increasing family $(\pi^\alpha, \mathcal{M}^\alpha, \mathcal{D}^\alpha)$ of regular induced extensions is majorized by a regular induced extension $(\pi, \mathcal{M}, \mathcal{D})$.

<u>Proof:</u> The first statement follows easily from the fact that the product of two injective mappings is again injective.

For the second statement define $\mathcal{D} = \bigcup_\alpha \mathcal{D}^\alpha$ and $\pi(x)f = \pi^\alpha(x)f$ if $f \in \mathcal{D}^\alpha$ and $\mathcal{M} = \bigcup_\alpha \mathcal{M}^\alpha$ where $\mathcal{M}^\alpha$ is naturally imbedded in $\mathcal{M}^\beta$ for $\beta > \alpha$. This means $b \in \mathcal{M}^\alpha$ is defined on $\mathcal{D}$ and is bounded by lemma II.5. Let $b \in (\pi, \mathcal{M}, \mathcal{D})'_w$, then $\rho(b) = 0$ implies by $\rho(b) = \rho^\alpha(\rho_\alpha(b))$, where ρ^α is the map from $B(\mathcal{H}(\mathcal{D}^\alpha)) \to B(\mathcal{H}(\mathcal{D}^o))$ and ρ_α the map from $B(\mathcal{H}(\mathcal{D})) \to B(\mathcal{H}(\mathcal{D}^\alpha))$, that $\rho_\alpha(b) = 0$ or $E_\alpha b \restriction \mathcal{H}(\mathcal{D}^\alpha) = 0$ Since $\bigcup_\alpha \mathcal{H}(\mathcal{D}^\alpha)$ is dense in $\mathcal{H}(\mathcal{D})$ it follows that $b = 0$ which implies the second statement.

This shows that the concept of regular induced extensions is very reasonable. If we can associate with it a set then we can use Zorn's Lemma. Therefore two problems remain, namely to associate a set with it and also to construct them. To answer these questions we assume that we have given a regular induced extension.

II.8. Lemma:

1.) Let $(\hat{\pi},\hat{\mathcal{M}},\hat{\mathcal{D}})$ be an induced extension of $(\pi,\mathcal{D})$ such that $\rho\restriction\mathcal{M}$ is injective.

Let $\mathcal{M} = \rho(\hat{\mathcal{M}}) \subset (\pi,\mathcal{D})'_w$

$\mathcal{K} = \rho(\hat{\mathcal{M}}^+)$

and define

$\phi : \mathcal{M}\times\mathcal{M} \to \mathcal{M}$ by

$\phi(m_1,m_2) = \rho(\rho^{-1}(m_1)\cdot\rho^{-1}(m_2))$ then

(i) $\mathcal{M}$ is a selfadjoint subset of $(\pi,\mathcal{D})'_w$

(ii) $\mathcal{K}$ is a convex cone with

(α) $\mathcal{K}$ generates $\mathcal{M}$

(β) $1 \in \mathcal{K}$

(γ) if $m \in \mathcal{K}$ then exist $0 \leq \lambda(m) < \infty$ such that $\lambda(m)\,1 - m \in \mathcal{K}$

(iii) ϕ has the properties:

a) ϕ is an abelian product on $\mathcal{M}$ i.e.

$\phi(m_1,m_2) = \phi(m_2,m_1)$

$\phi(m_1,\phi(m_2,m_3)) = \phi(\phi(m_1,m_2),\, m_3) := \phi(m_1,m_2,m_3)$

$\phi(1,m) = m;\ \phi(m_1,\, m_2+\chi m_3) = \phi(m_1,m_2) + \chi\phi\,(m_1,m_3)$

$\phi(m_1,m_2)^* = \phi(m_2^*,m_1^*)$

b) ϕ is positive i.e. for $p \in \mathcal{K}$, $m_i \in \mathcal{M}$ and $f_i \in \mathcal{D}$ we have

$$\sum_{ij}(f_i,\phi(m_i^*,p,m_j)\,f_j) \geq 0$$

2.) Let $\mathcal{M},\mathcal{K},\phi$ satisfy (i), (ii) and (iii) then there exists an induced extension such that

a) $\rho\restriction\mathcal{M}$ is injective

b) $\mathcal{M} = \rho(\hat{\mathcal{M}}^+)$

c) $\mathcal{K} \subset \rho(\hat{\mathcal{M}}^+)$

d) $\phi(m_1,m_2) = \rho(\rho^{-1}(m_1)\cdot\rho^{-1}(m_2))$

e) this extension is unique up to unitary equivalence

3.) Let $(\hat{\pi}^i, \hat{\mathcal{M}}^i, \hat{\mathcal{D}}^i)$ $i = 1,2$ be two induced extensions such that $\rho^i \restriction \mathcal{M}^i$ is injective then $(\hat{\pi}^1, \hat{\mathcal{M}}^1, \hat{\mathcal{D}}^1) < (\hat{\pi}^2, \hat{\mathcal{M}}^2, \hat{\mathcal{D}}^2)$ (after some unitary transformation) if and only if $\mathcal{M}^1 \subset \mathcal{M}^2$ and $\phi^1 = \phi^2 \restriction \mathcal{M}^1 \times \mathcal{M}^1$

Proof: The first part is a simple application of Lemma II.3. The existence of an induced extension is just as easy if we remark that by (iii) b) we have a scalar product on $\mathcal{M} \times \mathcal{D}$. If $\mathcal{N}$ is the null space under this scalar product then $\mathcal{D} = \mathcal{M} \times \mathcal{D} / \mathcal{N}$. The rest is only simple computation.

If we have two different induced extensions such that $\rho^1 \restriction \hat{\mathcal{M}}^1$ and $\rho^2 \restriction \hat{\mathcal{M}}^2$ are injective and such that ρ^i fulfill all the required properties then a simple computation shows that $U \sum_i \hat{m}_i^1 f_i = \sum \rho_2^{-1} \rho_1 \hat{m}_i^1 f_i$ defines a unitary operator mapping $\mathcal{H}(\mathcal{D}^1)$ into $\mathcal{H}(\mathcal{D}^2)$ which also has all other required properties. The verification of the last part is then straight forward.

This last lemma shows that the unitary equivalence classes of regular induced extensions form a semi-ordered set characterized by subsets $\mathcal{M}$, $\mathcal{K}$ and functions ϕ from $\mathcal{M} \times \mathcal{M} \to \mathcal{M}$. Therefore we get:

II.9. Corollary:
Let $(\pi, \mathcal{D})$ be a representation of a partial $*$ - algebra, then there exist maximal regular induced extensions.

It remains only to give the explicit construction of a regular induced extension.

II.10. Lemma:
Let $x \varepsilon (\pi, \mathcal{D})'_w$ with $0 \leq x \leq 1$ and $x \neq \lambda 1$.

Let $\mathcal{M} = \{\lambda x + \mu(1-x);\ \lambda, \mu \varepsilon \mathbb{C}\}$, $\mathcal{K} = \{\lambda x + \mu(1-x),\ \lambda, \mu \geq 0\}$

and $\phi(\lambda_1 x + \mu_1(1-x), \lambda_2 x + \mu_2(1-x)) = \lambda_1\lambda_2 x + \mu_1\mu_2(1-x)$

then the conditions of Lemma II.8. are satisfied. $\hat{\mathcal{M}}$ is generated by 1 and a projector e with $\rho(e) = x$.

II.11. Lemma:

Let x be as in the previous lemma, then the extension defined by x is regular, if and only if x is extremal in the weakly compact set

$$((\pi, \mathcal{D})'_w)^+_1 = \{x \in (\pi, \mathcal{D})'_w;\ 0 \leq x \leq 1\}$$

Proof: Let $0 \leq x \leq 1$ and x an extreme point of $((\pi, \mathcal{D})'_w)^+_1$, then extremality is equivalent to the following: The equations $0 \leq x + y \leq 1$ and $0 \leq x - y \leq 1$ implies $y = 0$, or equivalently x is extremal if and only if $y \in (\pi, \mathcal{D})'_w$ and $-x \leq y \leq x$ and $-(1-x) \leq y \leq (1-x)$ implies $y = 0$.

Let now $w \in (\hat{\pi}, \{e,1\}, \hat{\mathcal{D}})'_w$ with $w = w^*$ and $||w|| = 1$ and $\rho(w) = 0$ for some w or $\rho(ew) + \rho(1-e)w) = 0$.
Now $-e \leq e\,w \leq e$ and $-(1-e) \leq (1-e)w \leq 1-e$ implies $-x \leq \rho(ew) \leq x$ and $-(1-x) \leq \rho((1-e)w) \leq 1-x$.
From $\rho(ew) = -\rho((1-e)w)$ and extremality of x follows $\rho(ew) = \rho((1-e)w) = 0$. Since the extension $(\hat{\pi}, \hat{\mathcal{D}})$ is induced by e and (1-e) we have $\hat{\mathcal{D}} = e\,\mathcal{D} + (1-e)\,\mathcal{D}$. Hence we get

$$(ef_1+(1-e)g_1, w(ef_2+(1-e)g_2)) = (f_1, ew\, f_2) + (g_1, (1-e)wg_2)$$

$$= (f_1, \rho(ew)f_2) + (g_1, \rho((1-e)w)g_2) = 0. \quad \text{This shows } w = 0$$

Since ρ commutes with the involution it follows from this that $\rho \restriction (\hat{\pi}, \{e,1\}, \hat{\mathcal{D}})'_w$ is injective.

From this we get:

II.12. Theorem:

1) Every regular induced extension of $(\pi, \mathcal{D})$ is majorized by a maximal one.
2) A regular induced extension of $(\pi, \mathcal{D})$ is maximal if and only if
$$(\hat{\pi}, \hat{\mathcal{M}}, \hat{\mathcal{D}})'_w = (\hat{\pi}, \hat{\mathcal{M}}, \hat{\mathcal{D}})'_s = \hat{\mathcal{M}}$$
3) To every extremal $x \in \{(\pi, \mathcal{D})'_w\}^+_1$ there exists a maximal induced extension $(\hat{\pi}, \hat{\mathcal{M}}, \hat{\mathcal{D}})$ and a projection $e \in \hat{\mathcal{M}}$ such that $x = \rho(e)$.

4) Assume that the Hilbert-space $\mathcal{H}(\mathcal{D})$ is separable, then $\mathcal{H}(\hat{\mathcal{D}})$ is also separable.

Proof: Statements 1,2 and 3 are collections of the previous results. So only 4 needs some consideration. The unit ball $\hat{\mathcal{M}}_1$ is a weak compact set and is mapped by ρ into the weak compact set $\{(\pi, \mathcal{D})'_w\}_1$. Since ρ is continuous and injective it follows that ρ^{-1} is also continuous. Since $\mathcal{H}(\mathcal{D})$ is separable, it follows that the weak topology on $\{(\pi, \mathcal{D})'_w\}_1$ is countable and hence also the weak topology of $\mathcal{M}_1$. Since $\mathcal{M}_1 \mathcal{H}(\mathcal{D})$ is total in $\mathcal{H}(\hat{\mathcal{D}})$ it follows that $\mathcal{H}(\hat{\mathcal{D}})$ is separable.

III. Extensions with Positivity Conditions, Abelian Algebras

In many applications we want that certain elements of the partial algebra A are represented by positive operators. In order to handle also this situation, we introduce the following notations:

III.1. Definition:

1) Let A be a partial $*$ - algebra and A_h its hermitian part. A cone $P \subset A_h$ is called a regular cone if

   (i) $P \ni 1$

   (ii) if $x \in P$ and $y \in A$ and if $y^* xy$ is defined then follows $y^* xy \in P$

   (iii) $P \cap - P = 0$

2) Let P be a regular cone in A. A representation $(\pi, \mathcal{D})$ is called P-positive if $f \in \mathcal{D}$ and $x \in P$ implies $(f, \pi(x)f) \geq 0$.

3) Let $(\pi, \mathcal{D})$ be a P-positive representation of A. An extension $(\hat{\pi}, \hat{\mathcal{D}})$ is called a P-positive extension if $(\hat{\pi}, \hat{\mathcal{D}})$ is an extension of $(\pi, \mathcal{D})$ and $\hat{\pi}(x)$ is a positive operator for every $x \in P$.

If we have a P-positive representation of A then we can treat the extension theory in almost the same way as in the last section The only change consists in introducing a different order amongst operators in the weak commutant.

III.2. Definition:

1) Let $(\pi, \mathcal{D})$ be a P-positive representation and $x \in (\pi, \mathcal{D})'_w$ then we define $x >> 0$ if $(f, \pi(p)xf) \geq 0$ for all $f \in \mathcal{D}$ and all $p \in P$.

2) By $C_1^+(\pi, \mathcal{D}, P)$ we denote all $x \in (\pi, \mathcal{D})'_w$ with $0 \ll x \ll 1$ and by $C(\pi, \mathcal{D}, P)$ the linear span of $C_1^+(\pi, \mathcal{D}, P)$.

3) Let $(\hat{\pi}, \hat{\mathcal{M}}, \hat{\mathcal{D}})$ an induced P-positive extension then we define

$$C_1^+(\hat{\pi}, \hat{\mathcal{M}}, \hat{\mathcal{D}}, P) = C_1^+(\hat{\pi}, \hat{\mathcal{D}}, P) \cap \hat{\mathcal{M}}.$$

Replacing now the set $\{(\pi, \mathcal{D})_w'\}_1^+$ resp. $\{(\hat{\pi}, \hat{\mathcal{M}}, \hat{\mathcal{D}})_w'\}_1^+$ by the convex weakly compact sets $C_1^+(\pi, \mathcal{D}, P)$ resp. $C_1^+(\hat{\pi}, \hat{\mathcal{M}}, \hat{\mathcal{D}}, P)$ we can now proceed as in the last section. The outcome is the following

III.3. Theorem:

1) Every P-positive representation $(\pi, \mathcal{D})$ of A is majorized by a maximal regular induced P-positive extension.

2) A regular induced P-positive extension $(\hat{\pi}, \hat{\mathcal{M}}, \hat{\mathcal{D}})$ is maximal if and only if

$$C_1^+(\hat{\pi}, \hat{\mathcal{M}}, \hat{\mathcal{D}}, P) = \mathcal{M}_1^+.$$

The main field of application of this last theorem is the case of abelian algebras. If one has given a representation $(\pi, \mathcal{D})$ of an abelian $*$ - algebra A one usually wants to know whether one can find an extension $(\hat{\pi}, \hat{\mathcal{D}})$ such that all symmetric operators are essentially selfadjoint on $\hat{\mathcal{D}}$ and that there spectral projections commute. In order to formulate the results we need some notations

III.4. Definition:

1) Let A be an abelian $*$ - algebra, a representation $(\pi, \mathcal{D})$ is called standard (Powers [1]) if
   (i) If $x \in A_h$ than $\pi(x)$ is essentially selfadjoint on $\mathcal{D}$.
   (ii) for $x, y \in A_h$ the spectral projections of $\pi(x)$ and $\pi(y)$ commute.

2) Let $V \subset A_h$ be a linear subspace, then we say V generates A if for $x \in A$ exists a finite number of elements $v_1 \ldots v_n$ V and a polynomial P such that $x = P(v_1, v_2, \ldots v_n)$.

3) Denote by V^* the algebraic dual of V and let $Z \subset V^*$ be a subset then we denote by

$$P(Z) = \{P(v_1,\dots v_n);\ P(\omega(v_1),\omega(v_2),\dots\omega(v_n)) \geq 0$$
$$\text{for all } \omega \in Z\}.$$

with these notations we get the following result:

III.5. Theorem:
Let A be an abelian * - algebra generated by $V \subset A_h$ and Z a subset of V^*. Let π be a cyclic representation of A with cyclic vector Ω i.e. $\mathcal{D} = \pi(A)\Omega$. Then the following statements are equivalent:

1) $\Leftrightarrow$ 2) A) $\Leftrightarrow$ B)

1) The functional $T(x) = (\Omega,\pi(x)\Omega)$ is positive on P(Z)

2) The representation $(\pi,\pi(A)\Omega)$ is P(Z) positive and has a maximal regular induced P(Z)-positive extension

A) $(\hat{\pi},\hat{\mathcal{M}},\hat{\mathcal{D}})$ is a maximal regular induced P(Z)-positive extension of $(\pi,\pi(A)\Omega)$.

B) $(\hat{\pi},\hat{\mathcal{M}},\hat{\mathcal{D}})$ is a regular induced extension such that
   (a) $\hat{\mathcal{M}}\Omega$ is dense in $\mathcal{H}(\hat{\mathcal{D}})$
   and $\hat{\mathcal{M}}$ is a maximal abelian algebra.
   (b) $\hat{\pi}$ is standard
   (c) The joint spectrum of $\hat{\pi}(v_1)\dots\hat{\pi}(v_n)$, $v_i \in V$ belongs to the closure of the set
   $\{\omega(v_1),\omega(v_2),\dots\omega(v_n);\ \omega \in Z\}$
   (which is a subset of $\mathbb{R}^n$).

Proof: 1 $\Leftrightarrow$ 2 Since P(Z) is a regular cone it follows with $x \in P$ and $y \in A$ that also $y^* xy \in P$. Hence if T is positive on P(Z) then follows $(\pi,\pi(A)\Omega)$ is a P(Z)-positive representation. The second part of 2) is then just the last theorem. The converse conclusion is trivial.

B $\Rightarrow$ A. Since $\hat{\pi}$ is standard it follows from c) and the spectral theory that $(\hat{\pi},\hat{\mathcal{M}},\hat{\mathcal{D}})$ is P(Z) positive. Since $\hat{\mathcal{M}}$ is maximal abelian it follows that $C_1^+(\hat{\pi},\hat{\mathcal{M}},\hat{\mathcal{D}},P(Z)) = \hat{\mathcal{M}}_1^+$ which shows that

$(\hat{\pi}, \hat{\mathcal{M}}, \hat{\mathcal{D}})$ is a maximal regular induced (P(Z) positive extension.

A $\Rightarrow$ B. Let $Q(v_1, \ldots v_n)$ be a real polynomial. Let S be a finite dimensional linear subspace of $V + \hat{\mathcal{M}}_h$ such that $v_1, \ldots v_n \in S$. Denote by A_S the algebra generated by S and $\mathcal{H}_S = \overline{\hat{\pi}(A_S)\Omega}$. $T(x) = (\Omega, \hat{\pi}(x)\Omega)$ restricted to A_S defines a finite dimensional moment problem. Since the moments are $P_s = P(Z) \cap A_S$ positive, there exists a P positive solution defined on a Hilbert-space $\tilde{\mathcal{H}}_S \supset \mathcal{H}_S$. Denote by E_S the projection onto $\mathcal{H}_S$ and $\tilde{Q}_S$ be the representative of P in $\tilde{\mathcal{H}}_S$ which is selfadjoint. $(Q_s^2 + 1)^{-1}$ is a bounded operator such that $C = E_S\ (\tilde{Q}_S^2 + 1)^{-1} E_S$ is an element of $C_1^+\ (A_S, \mathcal{D}_S, P_S)$ and $C\ (\tilde{Q}_S^2 + 1)\Omega = \Omega$.

For every S denote by $\mathcal{L}_S$ the set of bounded operators on $\mathcal{H}(\hat{\mathcal{D}})$ such that

(i) $C\ (\tilde{Q}_S^2 + 1)\Omega = \Omega$

(ii) $||C|| \leq 1$

(iii) $C\ \mathcal{H}_S \subset \mathcal{H}_S$

(iv) $C \upharpoonright \mathcal{H}_S \in C_1^+\ (\hat{\pi}(A_S), \hat{\pi}(A_S)\Omega, P_S)$

By the above construction follows $\mathcal{L}_S \neq \emptyset$ and $\mathcal{L}_S$ is convex and weakly closed. For $S_1 \subset S_2$ follows $\mathcal{L}_{S_2} \subset \mathcal{L}_{S_1}$. This shows that the $\mathcal{L}_S$ have the finite intersection property. Since the unit ball is weakly compact follows $\bigcap_S \mathcal{L}_S \neq \emptyset$

If C belongs to this intersection then follows $C \in C_1^+\ (\hat{\pi}(A) \vee \hat{\mathcal{M}}, \hat{\mathcal{M}}, \hat{\mathcal{D}}, P(Z))$ and hence by maximality follows $C \in \hat{\mathcal{M}}_1^+$. From $C\ \hat{\pi}(Q^2+1)\Omega = \Omega$ follows $C = \hat{\pi}(Q^2+1)^{-1}$ showing that $\hat{\pi}(Q^2+1)$ is essentially selfadjoint on $\hat{\mathcal{D}}$ and is affiliated to $\hat{\mathcal{M}}$. Since now $\{\hat{\pi}(Q^2+1)\Omega, \hat{\mathcal{M}}\Omega\}$ span all of $\hat{\mathcal{D}}$ it follows that $\hat{\mathcal{D}}$ is maximal abelian. Hence $\hat{\pi}$ is standard.

Remark:

It is also possible to treat the noncyclic case. To do this one has to extend the notation of positivity. If this is done, then the result is similar to the last theorem.

IV. Integral Decomposition of States

In the following let us assume, that A is a $*$ - algebra. If we have a representation $(\pi, \mathcal{D})$ of A then we have seen, that we can construct maximal regular induced extensions $(\hat{\pi}, \hat{\mathcal{M}}, \hat{\mathcal{D}})$. Moreover, if $\mathcal{H}(\mathcal{D})$ was a separable Hilbert space then the same is true for $\mathcal{H}(\hat{\mathcal{D}})$. In this case we can make an integral decomposition of $\mathcal{H}(\hat{\mathcal{D}})$ with respect to $\hat{\mathcal{M}}$. This means there exists a locally compact space Λ and a finite positive Borel measure μ on it such that

$$\mathcal{H}(\hat{\mathcal{D}}) = \int_{\Lambda}^{\oplus} \mathcal{H}_{\lambda} \, d_{\mu}(\lambda)$$

and such that $\hat{\mathcal{M}}$ consists of all bounded diagonal operators with respect to this decomposition.

Now it is natural to ask whether one can decompose also the representation $\hat{\pi}(A)$. Since $\hat{\pi}(x)$ is generally an unbounded operator, such a decomposition is not always possible. The only case which I know of which yields integral decompositions beyond on Neumans theory is the nuclear spectral theorem [2,3]. Therefore we will make the following

IV.1. Assumptions

1) $\mathcal{D}$ is a nuclear vector space and the imbedding $\mathcal{D} \to \mathcal{H}(\mathcal{D})$ is continuous
2) A is a separable topological space, and π is a continuous representation
3) $\pi(A)\,\mathcal{D} \subset \mathcal{D}$ and the map $A \times \mathcal{D} \to \mathcal{D}$,

   $$(x,f) \to \pi(x)f$$

   is separately continuous as a map from $A \times \mathcal{D} \to \mathcal{H}(\mathcal{D})$

Under these conditions we get the following result:

IV.2. Theorem:

Assume the above assumptions and the integral decomposition of $\mathcal{H}(\hat{\mathcal{D}})$ with respect to $\hat{\mathcal{M}}$, then we get for almost all λ:

1) There exists a linear mapping $E_\lambda : \mathcal{D} \to \mathcal{H}_\lambda$ such that

   a) $\mathcal{D}_\lambda = E_\lambda \mathcal{D}$ is a nuclear space in the final topology continuously imbedded in $\mathcal{H}_\lambda$ and dense in $\mathcal{H}_\lambda$

   b) For all $f \in \mathcal{D}$; $\lambda \to E_\lambda f$ is a measurable field.

2) There exists a linear mapping $\hat{\pi}(x) \to \pi_\lambda(x)$ into the linear operators on $\mathcal{D}_\lambda$ such that

   a) $E_\lambda \hat{\pi}(x) f = \pi_\lambda(x) E_\lambda f$

   b) $x \to \pi_\lambda(x)$ is a $*$ - homomorphism and $\pi_\lambda(x)$ is a continuous representation when equipped with the final topology

   c) for $x \in A$ and $f \in \mathcal{D}$ the map $(x, f_\lambda) \to \pi_\lambda(x) f_\lambda$ is separately continuous.

3) If $g \in \mathcal{D}$ and $\lambda \to g_\lambda$ is any measurable field representing g then

   a) $g_\lambda \in \mathcal{D}_\lambda$

   b) $\lambda \to \pi_\lambda(x) g_\lambda$ is a measurable field and

$$\hat{\pi}(x) g = \int_\Lambda^\oplus \pi_\lambda(x) g_\lambda \, d\mu(\lambda)$$

4) If $(\pi, \mathcal{D})$ is a cyclic representation i.e. $\mathcal{D} = \pi(A)\Omega$ then $\mathcal{D}_\lambda = \pi_\lambda(A)\Omega_\lambda$ with $\Omega_\lambda = E_\lambda \Omega$

5) $(\pi_\lambda, \mathcal{D}_\lambda)'_W = \mathbb{C} \cdot 1$

6) If $(\pi, \mathcal{D})$ is cyclic and $\pi(x)\Omega = 0$ then follows $\pi_\lambda(x)\Omega_\lambda = 0$

Proof:

1) The existence of $E_\lambda : \mathcal{D} \to \mathcal{H}_\lambda$ such that $f = \int E_\lambda f \, d\mu(\lambda)$ is the nuclear spectral theorem. Since E_λ is continuous follows $\mathcal{D}_\lambda \cong \mathcal{D} / \mathrm{Ker}\, E_\lambda$ is a nuclear space. The density of $\mathcal{D}_\lambda$ in $\mathcal{H}_\lambda$ will follow from 3 a).

2) Let $f, g \in \mathcal{D}$, $x \in A$ and

$$T_\lambda(f,g,x) = (E_\lambda f, E_\lambda \pi(x)g) - (E_\lambda \pi(x)f, E_\lambda g)$$

then follows for every bounded μ-measurable function $m(\lambda)$

$$\int m(\lambda)\, T_\lambda(f,g,x)d\mu = (f, \hat{m}\hat{\pi}(x)g) - (\hat{\pi}(x)f, \hat{m}g) = 0$$

and hence

$$T_\lambda(f,g,x) = 0 \text{ a.e.}$$

Define $\pi_\lambda(x)E_\lambda\, g \ := E_\lambda \pi(x)g$,

then $\pi_\lambda(x)$ defines a $*$ - representation a.e. since $T_\lambda = 0$ and A is separable. The rest follows again since Ker E_λ is closed.

3) If $f \in \mathcal{D}$ then $f = \sum \hat{m}_i\, g_i$ with $\hat{m}_i \in \hat{\mathcal{M}}$ and $g_i\ \mathcal{D}$.
Since $\hat{\mathcal{M}}$ is diagonalisable follows $\hat{m}_i \to (\lambda \to m_i(\lambda)\cdot 1_\lambda)$ and hence f is represented by

$$\lambda \to \sum m_i(\lambda)\, E_\lambda g_i$$

which is measurable. Since $\hat{\mathcal{D}}$ is dense in $\mathcal{H}(\hat{\mathcal{D}})$ follows $E_\lambda\, \mathcal{D}$ is dense in $\mathcal{H}_\lambda$.

Now $\hat{\pi}(x)f = \hat{\pi}(x)\sum \hat{m}_i\, g_i = \sum \hat{m}_i \pi\,(x)\, g_i$

$$= \int \sum m_i(\lambda)\, \pi_\lambda(x)E_\lambda g_i d\mu(\lambda)$$

$$= \int \pi_\lambda(x) \sum m_i(\lambda)\, E_\lambda g_i d\mu(\lambda)$$

4) follows from 2 and so does 6.

5) The proof of this is exactly as in the case of von Neumann algebras which can be carried over since only matrix elements are needed in the proof.

Now we shall apply the last result to the integral decomposition of states. We will assume that A is a nuclear separable $*$ - algebra. The cases of physical interest are test function algebras over either $\mathcal{S}(\mathbb{R}^4)$ or $\mathcal{D}(\mathbb{R}^4)$. In both cases cases the algebra is separable. We will also assume that $1 \in A$ and that the product is separately continuous. Let now ω be a state on A, i.e. a continuous positive linear functional on A such that $\omega(1) = 1$.

By the G.N.S. construction we get a representation $(\pi_\omega, \pi_\omega \Omega)$.
In this case $\mathcal{D} = \pi_\omega \Omega$ is automatically a separable nuclear space which is continuously imbedded in $\mathcal{H}(\mathcal{D})$.

Furtheron we will assume that the map $A \times \mathcal{D} \to \mathcal{H}(\mathcal{D})$ defined by $\pi(x)f$ is separately continuous. Since π is generally only weakly continuous, it is a separate assumption. But in the case where A is a barrelled topological space (as in the cases of interest) then $x \to ||\pi(x)f||$ is automatically continuous (see e.g. [4]). With these assumptions we get:

IV.3. <u>Theorem:</u>
Let A be a nuclear * - algebra with separately continuous product, and ω a state on A. Assume that for every $y \in A$, $x \to \omega(y^* x^* x y)^{1/2}$ is continuous.

Then exists a locally compact space Λ and a positive normalized Borel measure $\mu(\lambda)$ on Λ and states ω_λ on A such that

1) $\omega = \int_\Lambda \omega_\lambda \, d\mu(\lambda)$ is a weak integral decomposition and μ-almost everywhere
2) ω_λ is an extremal state
3) $x \to \omega_\lambda(y^* x^* x y)^{1/2}$ is continuous
4) If $L_\omega = \{x \in A;\ \omega(x\ x) = 0\}$ then $L_\omega \subset L_{\omega_\lambda}$
5) $\pi_\omega(x)\Omega = \int_\Lambda \pi_{\omega_\lambda}(x)\Omega_\lambda d\mu(\lambda)$

   where π_ω and π_{ω_λ} fulfill the conditions of Theorem IV.2.

<u>Proof:</u>
Let $(\pi_\omega, \mathcal{D}_\omega)$ be the G.N.S. representation of ω then $\mathcal{D}_\omega \cong A/L_\omega$ and hence a separable nuclear space.
Then we construct some maximal regular induced extension $(\hat{\pi}_\omega, \hat{\mathcal{M}}, \hat{\mathcal{D}})$ and apply Theorem IV.2. to it. Define $\omega_\lambda(x) = (E_\lambda \Omega, E_\lambda\ \pi_\omega(x)\Omega) / (E_\lambda \Omega, E_\lambda \Omega)$ which gives the desired result.

V. The Moment Problem over Nuclear Spaces

We now want to apply our results to abelian $*$ - algebras. If A is an abelian $*$ - algebra of bounded operators then every extremal state is a character, it is a positive linear functional ω fulfilling

$$\omega(xy) = \omega(x)\omega(y)$$

But it is well known that on an algebra of unbounded operators not every extremal state is a character (non-solvability of the moment problem in more than one operator). Therefore the question arises to characterize those states which can be decomposed into characters.

V.1. Assumptions:

1) A is an abelian nuclear $*$ - algebra $1 \in A$ with a separately continuous product.
2) V is a linear subspace of A_h such that A(V) the algebra generated by $V \cup 1$ is dense in A.
   V is a real nuclear space in the topology induced by A.
3) V' denotes the (real) dual space of V and $Z \subset V'$ a subset P(Z) is the cone of polynomial $P(v_1 \ldots v_n)$ such that $P(\phi(v_1), \phi(v_2) \ldots \phi(v_n)) \geq 0$ for all $\phi \in Z$

With these notations we get:

V.2. Theorem:

Let A, V, Z as above and let ω be a state then the following statements are equivalent

1) ω is positive on P(Z) and $x \to \omega(x^* x)^{1/2}$ is continuous
2) ω has a weak integral decomposition

$$\omega = \int_\Lambda \omega_\lambda \, d\mu(\lambda)$$

where Λ is a locally compact space and μ is a positive normalized measure and the following holds for almost all

a) ω_λ is a character on A
b) $\omega_\lambda \restriction V$ belongs to the weak closure of Z.
c) $L_{\omega_\lambda} \subset L_\omega$
d) There exists a continuous seminorm p on A and a function $C \in L_2\ (\Lambda,\mu)$ with $C \geq 0$ and $|\omega_\lambda(x)| \leq C(\lambda)\ p\ (x)$

Proof:
The implication 2 $\Rightarrow$ 1 is straightforward. The other direction we get by combining the results of the sections III and IV.

In the physical applications there appear special cases of algebra, namely symmetric tensor algebras over nuclear spaces. Since there are generally more than one way of defining a tensor algebra we want to be sufficiently specific.

V.3. Assumptions:

1) Let V be real linear nuclear vector space and V' its dual. Assume V is the strict inductive limit of a countable number of its subspaces V_k

$$V = \varinjlim V_k$$

2) We define the nth tensorial power of V as

$$V^{\hat{\otimes} n} = \varinjlim V_k \mathbin{\hat{\otimes}} V_k \mathbin{\hat{\otimes}} \dots \mathbin{\hat{\otimes}} V_k$$

where $\hat{\otimes}$ denotes the complete π tensor product and

$$\underline{V} = \bigoplus_{n=o} (V^{\hat{\otimes} n} + i\ V^{\hat{\otimes} n})$$

equipped with the direct sum topology.

3) $S(V)$ the symmetric $\mathbb{C}$-tensor algebra which is derived from $\underline{V}$ in the standard way.

V.4. Lemma:

1) If ω is a continuous positive linear functional on S(V) then

$x \to \omega(x^* x)^{1/2}$ is continuous.

2) Let $Z \subset V'$, then the continuous characters on $S(V)$ which are positive on $P(Z)$ are in one to one correspondence with the elements of the weak closure Z via the formula

$$\omega(P(v_1,\ldots v_n)) = P(t(v_1),\ldots t(v_n))$$

$v_i \in V$ and $t = \omega \restriction V \in \bar{Z}$

The proof of these statements are fairly simple so that we can proceed in our investigation.

V.5. Theorem:

Let T be a linear functional on $S(V)$ then the following concitions are equivalent.

1) T is continuous and positive on $P(Z)$ with $T(1) = 1$

2) T has a weak integral decomposition

$$T(P(v_1,\ldots v_n)) = \int_\Lambda P(\omega_\lambda(v_1),\ldots \omega_\lambda(v_n))d\mu(\lambda)$$

where (Λ,μ) is a standard measure space with $\mu \geq 0$ and $\mu(\Lambda)=1$ and

a) ω_λ $\bar{Z}$

b) $\lambda \to \omega_\lambda(v)$ is μ measurable for all $v \in V$ and there exists a function $C(\lambda) \geq 0$, $C(\lambda)$ $\mathcal{L}_2(\Lambda,\mu)$ and continuous seminorms P_n on V with

$$|\omega_\lambda(v)| \leq C(\lambda)^{1/n} P_n(v), \quad n = 1,2,\ldots$$

3) $$T(P(v_1,v_2,\ldots v_n)) = \int_{\bar{Z}} P(\omega(v_n),\ldots \omega(v_n))\, d\nu_\omega$$

where ν is a measure on the σ-algebra generated by the weakly closed sets in V' and having the following property:

For any polynomially bounded continuous function f on R^n the integral

$$\int_{\bar{Z}} f(\omega(v_1),\ldots \omega(v_n))\, d\nu_\omega$$

exists and is jointly continuous in $v_1 \ldots v_2 \in V$.

Proof:
1) is implied by 2) or 3) in an obvious fashion. 1) implies 2) is theorem V.2. except for the estimate. But we have from theorem V.2.

$$|T_\lambda(v_1 \otimes v_2 \otimes \ldots \otimes v_n)| =$$

$$|\omega_\lambda(v_1) \cdot \omega_\lambda(v_2) \ldots \omega_\lambda(v_n)| \leq C(\lambda) p(v_1 \otimes v_2 \otimes \ldots \otimes v_n)$$

where p is a continuous seminorm on S(V). But there exists a continuous seminorm q_n on V such that

$$p \restriction V^{\otimes n} \leq q_n^{\otimes n} \quad \text{and therefore } |\omega_\lambda(v)| \leq c(\lambda)^{1n} q_n(v)$$

2) $\Rightarrow$ 3) By Lemma V.4. we have a one-valued map F from Λ into V'. Define a set M V' be measurable if F^{-1} (M) is measurable in Λ and $\nu(M) = \mu(F^{-1}(M))$. Measurable functions can be transported in an analogue manner. It remains to show that all weakly closed sets of V' are measurable. Let $K \subset V'$ be weakly closed then $\omega_\lambda \in K$ if and only if the character χ_λ defined by ω_λ is positive on the cone P(K). Using Lemma V.4. we see that all these characters are continuous with respect to a fixed seminorm p on S(V) and therefore P(K) is separable. If $P \in P(K)$ then the set $\{\lambda; \chi_\lambda(P) \geq 0\}$ is a μ-measurable and therefore also $\{\lambda; \omega_\lambda \in K\}$ is measurable as countable intersection of measurable sets.
The continuity property remains then a consequence of a simple estimate.

VI. Application to Quantum Field Theory

A Wightman functional W is usually defined as a state over the testfunction algebra $\mathcal{S}$ fulfilling the conditions (see e.g. [5])

(α) W is translational invariant

(β) W annihilates the two-sided locality ideal I_c

(γ) W annihilates the left spectral ideal

If $(\pi_W, \mathcal{D})$ is the cyclic representation constructed from W then due to the spectrum condition follows that also every operator b belonging to the weak commutant of this representation commutes automatically with the unitary representation of the translation group. From this follows:

VI.1. Lemma:
Let $(\hat{\pi}, \hat{\mathcal{M}}, \hat{\mathcal{D}})$ be a maximal regular extension of $(\pi_W, \mathcal{D})$ then the unitary group representation U(a) in $\mathcal{H}(\mathcal{D})$ of the translation group extends to a unitary group representation $\hat{U}(a)$ on $\mathcal{H}(\hat{\mathcal{D}})$ such that $\hat{\mathcal{M}}$ and $\hat{U}(a)$ commute.

As a consequence of this we get

VI.2. Theorem:
Every Wightman-state W can be decomposed into a weak integral over extremal Wightman states

$$W = \int_\Lambda W_\lambda d\mu(\lambda)$$

In the framework of local algebras of bounded operators it is well known that the extremality of the state and the uniqueness of the vacuum in the representation space are equivalent [6]. But, due to pathologies associated with unbounded operators it is possible to show that this is no longer true for Wightman fields. We now want to investigate under what conditions a

Wightman state can be decomposed into extremal ones with a unique vacuum. We start first with some

VI.3. Notations and Remarks

1) Let A(f) be some Wightman field ($f \in \mathcal{S}$) defined on some domain $\mathcal{D}$ in a Hilbert space $\mathcal{H}$ and let U(a) be the unitary representation of the translation group which is defined with it. Denote by P_o the projection onto the subspace invariant under U(a), and define $\mathcal{H}_o = P_o \mathcal{H}$.
2) Assume there is a cyclic subset $\mathcal{Z}_o \subset \mathcal{H}_o$ such that $\mathcal{D}$ = linear span $\{A(f)\, \mathcal{Z}_o\}$. Denote by $\bar{\mathcal{D}}$ the completion of $\mathcal{D}$ in the graph topology induced by all A(f). Define $\mathcal{J}_o = \bar{\mathcal{D}} \cap \mathcal{H}_o$ and $\mathcal{J}$ = linear span $\{A(f)\, \mathcal{J}_o\}$.
3) Due to the spectrum condition, the following statements are true (see [7])
   a) $P_o \mathcal{J} \subset \mathcal{J}_o$.
   b) The operators $A_o = P_o\, A\, P_o$ are well defined and generate a commutative $*$ - algebra on $\mathcal{J}_o$.
   c) $(A_o, \mathcal{J}_o)$ is a representation of $S(\mathcal{S})$ i.e. the symmetric tensor algebra over $\mathcal{S}$.
4) Let $\mathcal{S}^+$ be the set of positive elements in $\mathcal{S}$ i.e. $\left\{\sum_i f_i^+ f_i\ ;\right.$ the sum converges in $\left.\mathcal{S}\right\}$ and denote by $P(\mathcal{S}^+) \subset S(\mathcal{S})$ the set

$$\left\{\sum_i f_i^+ p_i f_i\ ;\ p_i \in \mathcal{S}^+\ ;\ f_i \in S(\mathcal{S})\right\}.$$

With this notation we have more precisely $(A_o, \mathcal{J}_o)$ is a $P(\mathcal{S}^+)$-positive representation of $S(\mathcal{S})$.

With these notations we get

VI.4. Theorem:

1) There is a one to one correspondence between
   1. unitary equivalence classes of induced regular extension $(\hat{A}, \hat{\mathcal{H}}, \mathcal{J})$ of $(A, \mathcal{J})$ and
   2. unitary equivalence classes of regular induced $P(\mathcal{S}^+)$-positive extensions

$(\hat{A}_o, \hat{\mathcal{M}}_o, \hat{\mathcal{D}}_o)$ of $(A_o, \mathcal{D}_o)$.

2) If $(\hat{A}, \hat{\mathcal{M}}, \hat{\mathcal{D}})$ and $(\hat{A}_o, \hat{\mathcal{M}}_o, \hat{\mathcal{D}}_o)$ are corresponding extensions then the following diagram commutes

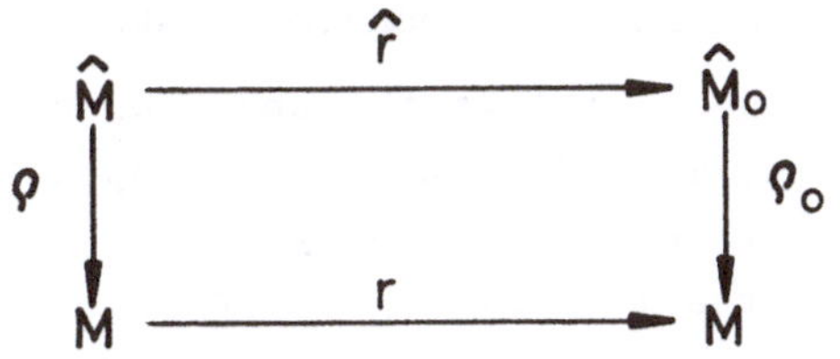

where the horizontal arrows stand for the restriction of invariant operators to the invariant subspace. It is a normal isomorphism of * - algebras since $\hat{\mathcal{M}}$ belongs to the strong commutant of $\hat{A}$. ρ and ρ_o are the weakly continuous mappings discussed earlier.

<u>Proof:</u>
The details of the proof are easy computations after having answered the following questions. Take a positive operator $b \in (A_o, \mathcal{D}_o)'_w$ when can it be lifted to a positive operator $\tilde{b}$ in $(A, \mathcal{D})'_w$ such that $b = P_o \tilde{b}$?

Assume $\tilde{b} \in (A, \mathcal{D})'_w$, $1 \geq \tilde{b} \geq 0$ and $f \in \mathcal{D}_o$.

then we have

$$0 \leq (A(x)f, \tilde{b} A(x)f) = (A(x x)f, \tilde{b}f) = (A(x x)f, P_o \tilde{b} f)$$

$$= (A_o(x x)f, P_o \tilde{b} f)$$

This shows $P_o \tilde{b} \in C_1^+ (A_o, \mathcal{D}_o, P(\mathcal{S}^+))$.

If on the other hand $b \in C_1^+ (A_o, \mathcal{D}_o, P(\mathcal{S}^+))$ then we can define $\tilde{b}$ by the same equation.

We now want to apply this result to Wightman functionals. To this end we need some

VI.5. Notations:

1) Let W be a Wightman state and $(A, A(\underline{\mathcal{S}})\Omega)$ the cyclic representation constructed from W.
   Denote by A_o the associated representation of $S(\underline{\mathcal{S}})$ and $W_o(x) = (\Omega, A_o(x)\Omega)$ the corresponding state on $S(\underline{\mathcal{S}})$.
2) For every $x \in \underline{\mathcal{S}}^+$ we want that the spectrum of $A_o(x)$ is the positive halfline. Therefore the common spectral set is the dual cone of $\underline{\mathcal{S}}^+$ which we denote by $\underline{\mathcal{S}}'^+$
   Then $P(\underline{\mathcal{S}}'^+)$ denotes the set of all polynomials

   $\{P(x_1,\ldots x_n);\ x_i \in \underline{\mathcal{S}}_h,$ and for all $\phi \in \underline{\mathcal{S}}'^+$ we have

   $P(\phi(x_1),\ldots,\phi(x_n)) \geq 0\}$

We now get:

VI.6. Theorem:

Let W be a Wightman state, then the following conditions are equivalent:

1) W has a weak integral decomposition

$$W = \int_\Lambda W_\lambda \, d\mu(\lambda)$$

   with Λ and μ as usual and W_λ has the cluster property a.e.
2) W_o has a weak integral decomposition

$$W_o = \int_\Lambda W_{o\lambda} \, d\mu(\lambda)$$

   with Λ, μ as above and $W_{o\lambda}$ is $P(\underline{\mathcal{S}}^+)$ positive character a.e.
3) W_o is positive on $P(\underline{\mathcal{S}}'^+)$.

The proof is only a collection of previous results.

There is another area of applications, these are the so-called Schwinger functionals and which are used extensively in the constructive field theory. But in order to treat the problems which are connected with them, some further results are needed which are beyond this representation given here. The results which we have obtained in Göttingen so far have been presented at the conference in Marseille.

References:

[1] POWERS, R.T., Comm. Math. Phys. 21, 85 (1971)

[2] MAURIN,K.: General Eigenfunction Expansions and Unitary Representation of Topological Groups. Warszawa Polish Scientific Publishers 1968

[3] GEL'FAND, I.M., VILENKIN, N.Ya.: Generalized Functions, Vol.4. New York, Academic Press 1964

[4] LASSNER,G.: Rep. Math. Phys. 3, 179 (1972)

[5] BORCHERS, H.J.: Algebraic Aspects of Wightman Field Theory. In:R.N.SEN and C.WEIL (Eds.) Statistical Mechanics and Field Theory (Haifa Summer School), New York, Halsted Press 1972

[6] ARAKI,H.: Progr. Theor. Phys. 32, 844 (1964)

[7] BORCHERS,H.J.: Comm. Math. Phys. 1, 49 (1965)

Acta Physica Austriaca, Suppl. XVI, 47–58 (1976)

Structure of (real-time) $P(\phi)_2$ Green's Functions

J. Dimock*
Département de Physique Théorique
Université de Genève

By Green's functions in the $P(\phi)_2$ model we shall mean vacuum expectation values of time-ordered products of field operators. The central role played by such objects in any many-body quantum mechanical problem is well-established. Here we discuss three general structural questions for Green's functions: first, how to compute derivatives with respect to coupling constants; second, how to take them apart by a kind of pull-through formula; and finally as an application of the first two, we sketch a proof that the perturbation series is asymptotic to all orders.

For simplicity we consider only a Hamiltonian of the form $H = H_o + \lambda \int :\phi(x)^d:dx - E$ with d an even integer, and we also assume that $\lambda \epsilon [o,\lambda_o]$ is sufficiently small so that the cluster expansion of Glimm, Jaffe, and Spencer converges [GJS]. The results undoubtedly have a much wider range of validity.

*On leave from SUNY at Buffalo.

A kind of generalized Green's function is defined as follows. Let θ be a non-negative complex number with $\operatorname{Im}\theta \leq o$, let $r=(r_1,\dots,r_n)$ be an n-tuple of non negative integers, and let $f=(f_1,\dots,f_n)$ be an n-tuple of functions $f_k \in C_o^\infty(R^2)$. Then we set

$$G_\theta(r,f)=\sum_\pi \int (\Omega, \prod_{k\in\pi} f_k(t_k,\vec{x}_k):\phi(\vec{x}_k)^{r_k}:e^{-i\theta H(t_k-t_{k'})}\Omega)d\vec{x}\, dt_\pi$$

where π is an ordering of $(1,\dots,n)$, k' is the predecessor of k in this ordering, and dt_π means integrate $(t_1,\dots,t_n)$ over the region respecting this ordering. A technique due to Nelson [N], [D2] together with the bound

$$\pm\int :\phi(\vec{x})^r:f(t,\vec{x})dx \leq C_f\lambda^{-1}(H+I)$$

valid for $r\leq d$, shows that $G_\theta(r,f)$ is well-defined, analytic in $\operatorname{Im}\theta<o$ and continuous in $\operatorname{Im}\theta\leq o$. Special cases are the real time Green's functions and the Schwinger functions defined respectively by

$$G(r,f) = G_{\theta=1}(r,f)$$

$$\hat{G}(r,f) = G_{\theta=-i}(r,f)$$

The Schwinger functions of course have a representation as the moments of a measure $d\nu$ on $\mathcal{S}'(R^2)$,

$$\hat{G}(r,f) = \int \prod_{k=1}^{n} :q^{r_k}:(f_k)\, d\nu$$

Formally $d\nu$ is given by

$$d\nu = \exp(-\lambda\textstyle\int :q(x)^d:dx)\, dq \Big/ \int \exp(-\lambda\textstyle\int :q(x)^d:dx)\, dq$$

where dq is the Gaussian measure with covariance $(-\Delta+m_o^2)^{-1}$.

I. Derivatives of G

From the path integral representation it can be shown [D1] that $\hat{G}(r,f)$ is a C^∞ function of λ and that the truncated Schwinger functions $\hat{G}^T(r,f)$ have derivatives given by

$$D_\lambda^m \hat{G}^T(r,f) = (-1)^m \sum_{I,J} \hat{G}^T(r,d;f,\chi_{I,J})$$

Here $I = (i_1,\ldots,i_m)$ and $J = (j_1,\ldots,j_m)$ are summed over Z^m and

$$\chi_{I,J} = (\chi_{i_1,j_1},\ldots,\chi_{i_m,j_m})$$

where $\chi_{i,j}$ is any suitable partition of unity, $\sum_{i,j} \chi_{i,j} = 1$, with time support near $i \varepsilon Z^1$, and space support near $j \varepsilon Z^1$. Due to the existence of a mass gap [GJS] the truncatea Schwinger functions are exponentially decreasing in all directions, and so the sum above converges.

A real time version of this result is contained in the following Theorem.

Theorem I [D2] There exists a partition of unity $\chi_{i,j}$ such that

$$|G_\theta^T(r,d;f,\chi_{I,J})| = O(\lambda^{-(n+m)}(|I|+1)^{-N}(|J|+1)^{-N})$$

for any N, uniformly in $\{\theta: \operatorname{Im}\theta \leq o, \frac{1}{2}<|\theta|<2\}$.In the same region $G_\theta(r,f)$ is C^∞ in λ on the interval (o,λ_o) and

$$D_\lambda^m G_\theta^T(r,f) = (-i\theta)^m \sum_{I,J} G_\theta^T(r,d;f,\chi_{I,J})$$

Sketch of Proof:

The derivative identities can be rewritten as integral identities, and once the bound is established the assertion is that two analytic functions with continuous boundary values are equal.

But the equality holds for $\theta=-i$ and by scaling for $\theta=-i\alpha,\alpha>o$. Thus it holds everywhere.

It is well-known that in real time theories with a mass gap there is exponential clustering in space-like directions. These bounds can be adapted to our case to show that in the region $|J|>|I|^2,C$ (some C) we have $|G_\theta^T(r,d;f,\chi_{I,J})| = O((|J|+1)^{-N})$ for any N. This bound is sufficient for this region.

In general we do not have rapid decrease for time like separation. It is here that we make a special choice for the partition of unity. We take $\chi_{i,j}(t,\vec{x}) = \rho_i(t)\,\chi_j(\vec{x})$ where χ_j is a standard partition and ρ_i satisfies

(a.) $\sum_i \rho_i = 1$

(b.) $||\rho_i||_1 \le K$

$||\rho_i^{(\alpha)}||_1 \le K_\alpha(|i|+1)^{-\alpha+1}$

(c.) $\operatorname{supp} \rho_i \subset \{t : |t-i|<\max(1,|i|/3m)\}$

Such a partition becomes smoother as $|i|\to\infty$. The ρ_i can be constructed as follows. Let ψ be a C^∞ function on $[o,\infty)$ satisfying $o \le \psi \le 1$, $\psi(t) = 1$ for $o \le t \le 1-1/3m$, and $\psi(t) = o$ for $t \ge 1$. For $i\epsilon Z^+$ we define $\psi_i(t) = \psi(t/i+1)$ and $\rho_i = \psi_i-\psi_{i-1}$. For i negative ρ_i is defined by reflection, and $\rho_o = \psi_o = \psi$. Then (a.), (b.),(c.) are satisfied.

Now we consider the other asymptotic region $|I|^2>|J|,C$; and we claim that $|G_\theta^T(r,d;f,\chi_{I,J})| = O((|I|+1)^{-N})$ for any N. For each configuration I we divide $(1,\dots,m)$ into two groups σ,σ' such that $d(\{o\}\cup I_\sigma, I_{\sigma'}) \ge |I|/m$, and for definiteness suppose $I_{\sigma'}$ comes first.

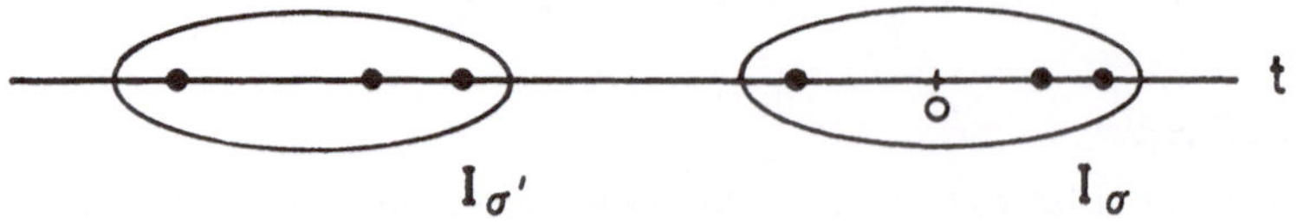

The corresponding functions then have supports separated in time and we may always isolate in $G_\theta^T(r,d;f,\chi_{I,J})$ a structure like

$$|| E_o^\perp \int e^{-i\theta H(T-t_{\underline{k}})} \prod_{k\epsilon\sigma'} \rho_{i_k}(t_k)\ \chi_{j_k}(\vec{x}_k)\ :\ \phi(\vec{x}_k)^d\ :$$

$$e^{-i\theta H(t_k - t_{k'})}\ \Omega\ d\vec{x}\ dt_{\sigma'}\ ||$$

where σ' has a definite ordering and $\underline{k}$ is the last element in σ'. On the orthogonal complement of the vacuum we may insert $(i\theta H)^{-N}(\partial/\partial t_{\underline{k}})^{-N}$. After integrating by parts the derivatives all appear on the $\{\rho_{i_k}\}_{k\epsilon\sigma'}$ and due to the property (b.) for the ρ_i the above expression is $O((|I|+1)^{-N+|\sigma'|})$ for any N. The desired bound follows.

The factor $\lambda^{-(n+m)}$ arises from the $:\phi^r:$ bounds which are used everywhere. This prevents a straightforward extension to $\lambda=o$. We return to this question in §III.

II. Expansion for G.

The problem we now consider is how to take the Green's functions apart by making a finite expansion in powers of the interaction. Experience in constructive quantum field has shown this to be a valuable technique for analyzing the structure of such objects.

We begin again at imaginary time. We take the path space representation for $\hat{G}(r,f)$ and successively remove each linear factor of the fields by integration by parts using the formula

$$\int f(x):q(x)^r:R\, dxd\nu = \int f(x):q(x)^{r-1}:C(x-y)$$

$$(\delta R/\delta q(y) - \lambda d:q(y)^{d-1}:R)\, dxdyd\nu$$

where $C(x-y)$ is the kernel of $(-\Delta+m_o^2)^{-1}$. This yields [GJ]

$$\hat{G}(r,f) = \sum_{\Gamma} \int \prod_{v\in U_1} f_v(x_v) \prod_{\ell\in\mathcal{L}} C(x_{\ell_1}-x_{\ell_2}) \prod_{v\in U_2} -\lambda : q(x_v)^{r_v}:dxd\nu$$

Here the sum is over all graphs $\Gamma=(U,\mathcal{L})$ with vertices U, lines $\mathcal{L}$, and r_v = number of uncontracted legs at v. The vertices $U=U_1\cup U_2$ consists of intial vertices $U_1=U_1(r)$ which are fully contracted and new interaction vertices U_2 which are joined to the U_1 vertices. A typical graph might be:

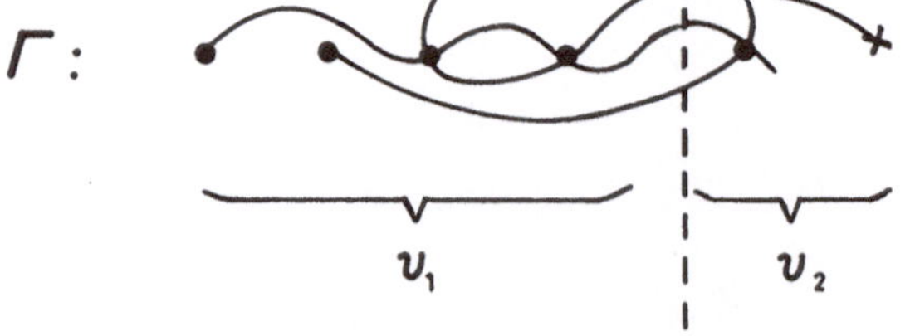

We introduce a partition of unity χ_{i_v,j_v} for each variable x_v, $v\varepsilon U_2$. Then with $I = \{i_v\}_{v\varepsilon U_2}$, $J = \{j_v\}_{v\varepsilon U_2}$ we have

$$\hat{G}(r,f) = \sum_{\Gamma}\sum_{I,J} \hat{I}(\Gamma;f,-\lambda\chi_{I,J})$$

where $f = \{f_v\}_{v\varepsilon U_2}, \chi_{I,J} = \{\chi_{i_v,j_v}\}_{v\varepsilon U_2}$, and where for general vertex functions $h = \{h_v\}_{v\varepsilon U}$ we define

$$\hat{I}(\Gamma,h) = \int \prod_{v\varepsilon U} h_v(x_v) : q(x_v)^{r_v} : \prod_{\ell\varepsilon\mathcal{L}} C(x_{\ell_1}-x_{\ell_2})\, dx d\nu$$

Note that due to the rapid decrease of C, $\hat{I}(\Gamma,f,\chi_{I,J})$ is rapidly decreasing in $|I|,|J|$ and so the sum converges.

To analytically continue we define

$$\Delta_\theta(x-x') = \int \exp\,(-i\,\vec{p}\cdot(\vec{x}-\vec{x}') - i\theta\omega|t-t'|)(4\pi\omega)^{-1}d\vec{p}$$

with $\omega=\sqrt{\vec{p}^2+m_o^2}$. Then $\Delta_{\theta=-i} = C$ and $\Delta_{\theta=1} = \Delta_F$, the Feynman propagator. We also define $I_\theta(\Gamma,h)$ so that $I_{\theta=-i}(\Gamma,h) = \hat{I}(\Gamma,h)$. Thus we have

$$\begin{aligned} & I_\theta(\Gamma,h) \\ &= \sum_V \int (\Omega, \prod_{v\varepsilon V} h_v(x_v) : \phi(\vec{x}_v)^{r_v} : e^{-i\theta H(t_v-t_{v'})}\Omega) \\ &\qquad \prod_\ell \Delta_\theta(x_{\ell_1}-x_{\ell_2})\, d\vec{x}\, dt_V \\ &= \sum_V \int (\Omega, \prod_{v\varepsilon V} e^{-i\vec{p}_v\cdot\vec{x}_v} h_v(x_v) : \phi(\vec{x}_v)^{r_v} : e^{-i\theta(H+\Omega_{v'})(t_v-t_{v'})}\Omega) \\ &\qquad d\vec{x}\, dt_V \prod_\ell (4\pi\omega_\ell)^{-1}d\vec{p}_\ell \end{aligned}$$

where the sum is over the orderings V of U. For each V, v' is the predecessor of v, dt_V is the integral over the region respecting the ordering, and furthermore

$$\vec{p}_v = \sum_{\ell\varepsilon\mathcal{L}_v^-} \vec{p}_\ell - \sum_{\ell\varepsilon\mathcal{L}_v^+} \vec{p}_\ell$$

$$\Omega_v = \sum_{\ell \in \mathcal{m}_v} \omega_\ell$$

where $\mathcal{L}_v^+$ ($\mathcal{L}_v^-$) is all lines joining v to a higher (lower) vertex, and $\mathcal{m}_v$ is all lines joining a vertex $v_1 \leq v$ to a vertex $v_2 > v$. The first expression is only formal, but the second can be made exact by a modification of Nelson's technique. The crucial points are that $\Omega_v \geq o$, and that each line connects at least one vertex which is fully contracted. Furthermore one shows that $I_\theta(\Gamma,h)$ is analytic in $\mathrm{Im}\theta < o$ and continuous in $\mathrm{Im}\theta \leq o$.

Theorem II [D3] There exists a partition of unity $\chi_{i,j}$ such that

$$|I_\theta(\Gamma;f,\lambda\chi_{I,J})| = O((|I|+1)^{-N}(|J|+1)^{-N})$$

for any N, uniformly in $\lambda\varepsilon(o,\lambda_o)$ and $\{\theta:\mathrm{Im}\theta<o, \frac{1}{2}<|\theta|<2\}$. In the same region:

$$G_\theta(r,f) = \sum_\Gamma \sum_{I,J} I_\theta(\Gamma;f,-i\theta\lambda\chi_{I,J})$$

Sketch of Proof:

Once the bound is established, we are asserting the identity of two analytic functions with continuous boundary values. But equality if known for $\theta=-i$ and by scaling also for $\theta=-i\alpha$, $\alpha>o$. Thus the identity holds.

The propagator Δ_θ is rapidly decreasing in space-like directions which enables one to show that in the region $|J|>|I|^2$, C we have $|I_\theta(\Gamma;f,\lambda\chi_{I,J})| = O(|J|+1)^{-N}$. For the other asymptotic region $|I|^2>|J|$, C we take $\chi_{i,j}$ as before and claim that $|I_\theta(\Gamma;f,\lambda\chi_{I,J})| = O(|I|+1)^{-N}$. To see this divide U_2 into two groups σ,σ' so $d(\{o\}UI_\sigma,I_{\sigma'}) \geq |I|/|U_2|$. Then we can isolate in $I_\theta(\Gamma;f,\lambda\chi_{I,J})$ a structure like

$$|| \int e^{-i\theta(H+\Omega_{\bar v})(T-t_{\bar v})} \prod_{v\varepsilon\sigma'} e^{-i\vec{p}_v\cdot\vec{x}_v} \rho_{i_v}(t_v)\chi_{j_v}(\vec{x}_v)$$

$$:\phi(\vec{x}_v)^{r_v} : e^{-i\theta(H+\Omega_{v'})(t_v-t_{v'})}\Omega \; d\vec{x}dt_{\sigma'}||$$

where σ' has a definite order, and $\bar{v}$ is the last element. Since σ' is joined to U_1 we have $\Omega_{\bar{v}} \geq m_o$ and so can insert $(i\theta)^{-N}(H+\Omega_{\bar{v}})^{-N}(\partial/\partial t_{\bar{v}})^N$ in the front. The derivatives end up on the $\{\rho_{i_v}\}_{v\epsilon\sigma'}$ and a bound $O(|I|+1)^{-N+|\sigma'|}$ follows as before

The uniformity in λ can be understood since all $:\phi^r:$ operators are multiplied by λ and $\lambda:\phi^r:$ is uniformly dominated by H.

Remark: Our expansion can be formally generated by a pull-through formula. That is we write

$$:\phi(\vec{x})^r: = \int (4\pi\omega)^{-1/2}\, d\vec{p}\, e^{-i\vec{p}\cdot\vec{x}}\, (a(\vec{p})^*:\phi(\vec{x})^{r-1}:+ \\ :\phi(\vec{x})^{r-1}:a(-\vec{p}))$$

and commute $a(\vec{p})^*$ and $a(-\vec{p})$ to the outside using

$$[a(\vec{p}),:\phi(\vec{x})^r:] = e^{-i\vec{p}\cdot\vec{x}}(4\pi\omega)^{-1/2} r:\phi(\vec{x})^{r-1}:$$

and

$$a(\vec{p})e^{-iH(t-s)} = e^{-i(H+\omega)(t-s)}a(\vec{p})$$

$$+\int_s^t e^{-i(H+\omega)(t-u)}\, [a(\vec{p}),\, -i\lambda \int :\phi(\vec{x})^d:d\vec{x}]\, e^{-iH(u-s)}du$$

The vacuum is considered to be

$$\Omega = \lim_{t\to\pm\infty} e^{iHt}\Omega_o .$$

Combining these elements gives the $\theta=1$ identity.

III. Perturbation Theory for G

The methods we have developed provide a proof that the perturbation series for G is asymptotic. Our use of Hamiltonians throughout provides a more elaborate structure than do the methods of Eckmann, Epstein, and Fröhlich [EEF] or Osterwalder and Sénéor [OS] who have similar results. (See these proceedings.) This is both good and bad.

As noted, the methods of §I do not allow us to conclude that G is C^∞ up to $\lambda=o$, which would imply that perturbation theory is asymptotic. If we continued the expansion of §II we would also generate the perturbation series, but it is hard to show that the remainder is small. Instead we use the expansion of §II to get uniform bounds on the derivatives as computed in §I. In particular we make the expansion

$$G^T(r,d;f,\chi_{I,J}) = \sum_{\Gamma} \sum_{I',J'} I^T(\Gamma;f,\chi_{I,J};-i\lambda\chi_{I',J'})$$

An expansion of this type for truncated Green's functions is valid when I^T is defined in the standard way inductively from I with the connected components of Γ as the basic units. [GJ], [D3].

Theorem III [D3]

(a.) Uniformly in $\lambda\varepsilon(o,\lambda_o)$, for any N,

$$|I^T(\Gamma;f,\chi_{I,J};\ \lambda\chi_{I',J'})| = O((|I|+1)^{-N}(|J|+1)^{-N}(|I'|+1)^{-N}(|J'|+1)^{-N})$$

$$|G^T(r,d;f,\chi_{I,J})| = O((|I|+1)^{-N}(|J|+1)^{-N})$$

$$|D_\lambda^m G^T(r,f)| = O(1)$$

(b.) $G^T(r,f)$ is C^∞ on $[o,\lambda_o)$. The derivatives $D_\lambda^m G^T(r,f)|_{\lambda=o}$ are the coefficients of the standard perturbation series. Hence the series is asymptotic.

Sketch of Proof:

Part (b.) is relatively easy once we know $|D_\lambda^m G^T(r,f)|$ is bounded in λ. All bounds in (a.) follow from the first. This is proved by isolating widely separated clusters of (o,I,I') (if time-like separation predominates) or of (o,J,J') (if space like separation predominates). If Γ connects the clusters the methods of §II give the bound, while if Γ does not connect the clusters the methods of §I give the bound.

References

[D1] J.DIMOCK, Commun. Math. Phys. 35, (1974), 347-356.

[D2] J.DIMOCK, The $P(\phi)_2$ Green's Functions: Smoothness in the Coupling Constant, to appear in J. Func. Anal.

[D3] J.DIMOCK, The $P(\phi)_2$ Green's Functions: Asymptotic Perturbation Expansion, to appear in Helv. Phys. Acta.

[EEF] J.-P.ECKMANN, H.EPSTEIN, J.FRÖHLICH, Asymptotic Perturbation Expansion for the S-matrix and the Definition of Time Ordered Functions in Relativistic Quantum Field Models, preprint.

[GJS] J.GLIMM, A.JAFFE, and T.SPENCER, Annals of Math. 100, (1974), 585-632.

[GJ] J.GLIMM, A.JAFFE, Two and Three Body equations in Quantum Field Models, to appear Commun. Math. Phys.

[N] E.NELSON, J.Func. Anal. 11, (1972), 211-219.

[OS] K.OSTERWALDER, R.SENEOR, The Scattering Matrix is Nontrivial for Weakly Coupled $P(\phi)_2$ Models, preprint.

Acta Physica Austriaca, Suppl. XVI, 59–68 (1976)

Asymptotic Perturbation Expansion for the S-matrix in $P(\phi)_2$ Quantum Field Theory Models

Jean-Pierre Eckmann
Département de Physique Théorique
Université de Genève

This talk describes results of a paper with the same title done jointly with H. Epstein and J. Fröhlich. Similar work has been done simultaneously and independently by J. Dimock, K. Osterwalder, and R. Sénéor, and since they also report at this conference I will not describe any of their methods.

I. Discussion of General Axiomatic Properties

The point of view we want to take is to derive properties of the scattering matrix in relativistic quantum field theory from properties of Schwinger functions (Euclidean Green's functions). To fix the notations, let me first introduce the objects we shall talk about. We consider most of the time, for convenience of exposition, a Wightman theory (with isolated mass m) depending only on the time variables, called nevertheless x_i.

Let $\omega^T(z)$ denote the analytic, truncated Wightman function, analytic in $\bigcup_\pi \mathcal{T}_\pi$, π the permutations of $\{1,..n\}$,

$$\mathcal{T}_\pi = \{z = x+iy \ ; \ y_{\pi_1} < y_{\pi_2} < \ldots < y_{\pi_n}\} \ , \ \text{(the permuted tube)}$$

with boundary value $\omega^T_\pi(x)$ in $\mathcal{T}_\pi$. At Euclidean points we have a function

$$\overset{o}{S}{}^T(y_1,\ldots y_n) = \omega(iy_1,\ldots iy_n), \tag{1}$$

the truncated Schwinger function (at non coinciding points!)
Given π, let $\zeta_j^\pi = z_{\pi_j} - z_{\pi(j+1)} = \xi_j^\pi + i\eta_j^\pi$, $j = 1,\ldots n-1$.
Suppose that time ordered functions

$$\tilde{h}_\pi(x_1,\ldots x_n) = \prod_{j=1}^{n-1} \Theta(\xi_j^\pi) \ \omega_\pi^T(x_1,\ldots x_n)$$

can be defined and let $\delta(\Sigma p_j)h_\pi(p)$ be their Fourier transforms. Set $h(p) = \sum_\pi h_\pi(p)$. Suppose furthermore the theory has a positive mass m. Then the S-matrix elements (if they exist) are given by

$$< p_1^{in} \cdot \cdot \cdot p_k^{in} \ p_{k+1}^{out} \cdot \cdot \cdot p_n^{out} >$$

$$= Z^{-n} \prod_{j=1}^{n} (p_j^2 - m^2) h(p) \Big|_{p_j^2 = m^2 \ \forall j} \tag{2}$$

by the LSZ reduction formula. Here, Z is the field strength (re)normalization and the variables have space and time components.

If $\theta\omega_\pi^T$ can be defined, then the following situation is known from axiomatic field theory (Bros, Epstein, Glaser, Ruelle, Stora)

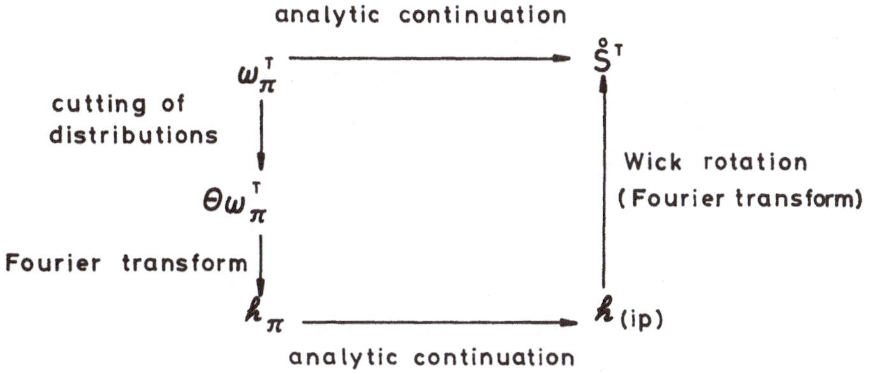

The Osterwalder-Schrader axioms, of which I only mention the two most important for our purposes imply an arrow $\omega_\pi^T \longleftarrow \mathring{S}^T$. They are

(I) $\Sigma a_n \bar{a}_m \mathring{S}_{n+m}(-\overleftarrow{t}_n, t_m) \geq 0$

for $a_j \in \mathbb{C}$, $t_m = \{t_1^{(m)} \dots t_m^{(m)}\}$, $t_1^{(m)} < t_2^{(m)} < \dots < t_m^{(m)}$,

$\overleftarrow{t}_m$ is t_m in inverse order,

and (II) $|\int \mathring{S}_n(t) f(t) dt| \leq (n!)^K ||f||_{\mathcal{S}}$, some Schwartz norm of order $\leq \mathcal{O}(n)$.

So this shows that $\mathring{S}^T$ can be given as the basic object. We now replace (II) by the stronger condition

(II') $|\overset{\circ}{S}_n(t)| \leq (n!)^K$ const. *

and we define S(t) to be the extension of $\overset{\circ}{S}(t)$ to coinciding points (as an integrable function).

Theorem 1:

a) (I) + (II') imply that $\Theta\omega_\pi^T$ can be properly defined.

b) (I) + (II') + existence of positive mass imply h(ip) is the Fourier transform of S^T.

So we have now the diagram:

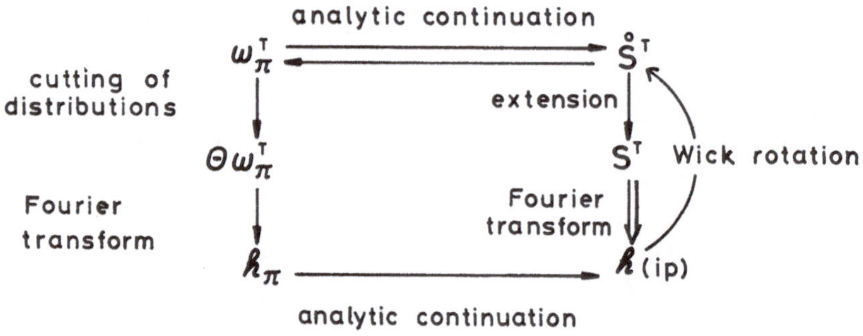

Discussion: At first sight, b) might look trivial. However suppose I am given S^T such that ⇓ holds. Then I might change S^T in such a way at coinciding points that h(ip) of this new S^T would not have an analytic continuation. So the content of b) could also be stated: The Fourier transform of S^T has an analytic continuation in the axiomatic domain.

Sketch of the main ideas of the proof:

We consider only the permutation $\pi=1$ and the tube $\mathcal{T}=\mathcal{T}_\pi$. Let always $\phi\in\mathcal{S}(\mathbb{R}^{|J'|})$, J, J' two complementary subsets of {1,...n-1} and $\zeta_J = \{\zeta_j\}_{j\in J}$.

---

* In our paper, this is replaced by a weaker condition of local integrability in the time-differences. We discuss here only the simpler assumption (II') and this has the effect of making notation shorter.

Lemma 1: For $\eta_j < o$, $j\varepsilon J$,

$$W(\phi, i\eta_J) \; := \; \lim_{\substack{\eta_j \uparrow o \\ j\varepsilon J'}} \int \omega(z)\phi(\xi_{J'})d\xi_{J'}\Big|_{z_J = i\eta_J}$$

exists and has an analytic continuation in the J-variables and

$$|W(\phi,\zeta_J)| \leq (1 + \sum_{j\varepsilon J} |\zeta_j|)^L (\sup_{r\varepsilon J, \xi_r \neq o} (-\eta_r))^{-N} ||\phi||_N \; , \qquad (3)$$

for a $\mathcal{S}$-norm $||\;||_N$ of order $\leq N$.

This lemma is a straightforward modification of the Glaser - Osterwalder - Schrader method of analytic continuation.

We now define $\theta\omega$. Let α be a function analytic in $\mathrm{Re}\zeta > -1$, $\mathrm{Im}\zeta < o$ and satisfying $\alpha(o) = 1$, $(\partial^r_\zeta \alpha)(o) = o$ for $r=1,\ldots N+1$ and sufficiently decreasing at infinity. The following lemma gives a precise construction of $\theta\omega$.

Lemma 2: The following limits exist:

$$\int \prod_{j\varepsilon J'} \theta(\xi_j)\omega(\xi_{J'}, \zeta_J)\phi(\xi_{J'})d\xi_{J'} \qquad (4)$$

$$= \lim_{\varepsilon \downarrow o} \lim_{\substack{\eta_j \uparrow o \\ j\varepsilon J'}} \int \omega(\zeta_{J'},\zeta_J) \prod_{j\varepsilon J'} (\theta(\xi_j)(1-\alpha(\xi_j/\varepsilon))\phi(\xi_{J'})d\xi_{J'} .$$

They satisfy:

(1) Independence of α.

(2) Support in the complement of $\xi_j < o$, $j\varepsilon J'$ (if all variables are considered, they satisfy correct locality properties).

(3) They are analytic in ζ_J in the region $\mathrm{Im}\zeta_j < o$, $j\varepsilon J$ and satisfy a bound of the form of (3) in Lemma 1.

Proof: Let u be an analytic function of the same type as α. We sketch the argument in one variable only. For $\phi \varepsilon \mathcal{S}(\mathbb{R})$, write

$$\phi = K_u\phi + (1-K_u)\phi$$

where

$$(K_u\phi)(\xi) = \sum_{k=o}^{N+1} u(\xi)\frac{\xi^k}{k!} (\partial^k_\xi \phi)(o) .$$

Then $\theta(\xi)(1-K_u)\phi(\xi)$ is an admissible test function for ω since N+1 derivatives vanish at the origin, and a direct calculation shows $\lim_{\varepsilon\to o}$ exists. On the other hand, $K_u\phi$ is an analytic test function and we can deform the integration contour, so that

$$\lim_{\varepsilon\to o}\int_{-\infty}^{+\infty}\omega(\xi)\theta(\xi)(1-\alpha(\xi/\varepsilon))(K_u\phi)(\xi)d\xi$$

$$=\lim_{\varepsilon\to o}\int_o^{-\infty}\omega(i\eta)(1-\alpha(i\eta/\varepsilon))(K_u\phi)(i\eta)id\eta =$$

$$\int_o^{-\infty}\omega(i\eta)(K_u\phi)(i\eta)id\eta .$$

The bounds can be read off the construction.

The other assertions are shown by similar calculations. We have now "proved" Lemma 2 and Theorem 1a).

Assume now the theory has a positive mass $m > o$, and consider the function

$$h(p) = (2\pi)^{-(n-1)}\sum_{\pi}\int\prod_{j=1}^{n-1}\theta(\xi_j^\pi)\omega_\pi^T(x)\exp i(\Sigma P_j^\pi\xi_j^\pi)d\xi_1^\pi..d\xi_{n-1}^\pi ,$$

where P_j is defined by $\sum_{j=1}^{n-1}P_j^\pi\xi_j^\pi = \sum_{k=1}^{n}p_k x_k,\ \sum_{k=1}^{n}p_k = o.$

Since $m > o$, $\tilde{\omega}_\pi^T(p)$ has support in $P_j^\pi \geq m$, $j = 1,...n-1$. From this it is easy to see that $h(p)$ is analytic in $\{p : \sum_{j\varepsilon J}p_j \notin m+\mathbb{R}^+,\ J\subset\{1,...n\}\}$. In particular $h_\pi(ip)$ is defined and equal to

$$h_\pi(ip) = (2\pi)^{-(n-1)}\int d\xi^\pi\prod_{j=1}^{n-1}\theta(\xi_j)\omega_\pi^T(x)\exp -\Sigma P_j^\pi\xi_j^\pi .$$

The claim of Theorem 1b) is that this can be written as the Fourier transform of $\omega^T(i\eta)$. Indeed, by Lemma 2, $B^A/(B+\xi)^A \times \exp - P\xi$ is an admissible test function for $\theta\omega$ and hence, since $m > o$ (convergence at infinity),

$$h_\pi(ip) = \lim_{B\to\infty}(2\pi)^{1-n}\int_{\eta_j^\pi<o}\omega^T(i\eta)\prod_k\left\{\frac{B^A}{(B+i\eta_k^\pi)^A}e^{-iP_k^\pi\eta_k^\pi}(-id\eta_k^\pi)\right\}$$

and this proves that $h(ip)=\Sigma_\pi h_\pi(ip)$ is the Fourier transform of S^T.

II. Application to $P(\phi)_2$

Theorem 1 provides us now with the necessary machinery to analyze differentiability of theories with respect to the coupling constant. (Our results are also valid for the cos ϕ_2 models). So consider the family of models depending on λ, m_o; λ/m_o^2 small, $\lambda \geq o$, m_o large, whose Schwinger functions are given by

$$S_{\underline{\nu}}(x_1,\dots x_n;m_o,\lambda)=\lim_{L\to\infty}\frac{\int :\phi^{\nu_1}(x_1):\dots:\phi^{\nu_n}(x_n):\exp(-\lambda\int_{|x|\leq L}:P(\phi):(x)dx)d\phi_{m_o}}{\int \exp(-\lambda\int_{|x|\leq L}:P(\phi):(x)dx)\; d\phi_{m_o}}\;.$$

Here P is a polynomial which is lower bounded, and $d\phi_{m_o}$ is the Gaussian measure on $\mathcal{S}'(\mathbb{R}^2)$ with mean zero and covariance $(-\Delta+m_o^2)^{-1}$.

We summarize facts about $S_{\underline{\nu}}$, sup ν_i bounded, in the following theorem, where we always make the assumption λ/m_o^2 small, $\lambda \geq o$.

Theorem 2: a) $S_{\underline{\nu}}$ satisfies the Osterwalder Schrader axioms (Glimm, Jaffe, Spencer).

b) S_ν is C^∞ in λ, m_o jointly (Dimock, Eckmann, Epstein, Fröhlich). The derivatives are given (e.g. ∂_λ and $P(\phi) = \phi^4$) by

$$\partial_\lambda S^T_{\underline{\nu}}(x_1,\dots x_n;m_o,\lambda) = -\int d^2y S^T_{\underline{\nu}\,\nu 4}(x_1,\dots x_n,y;m_o,\lambda).$$

c) $S_{\underline{\nu}}$, when smeared in space, is bounded in time by $(n!)^{\sup \nu_j}$ (EEF).

d) The theory has a mass-shell which is isolated from the remainder of the spectrum uniformly in the above (λ, m_o) region (Glimm, Jaffe, Spencer)

(Remark: c) could also be read off from work of Fröhlich, Glimm, Jaffe, Spencer). The proofs of b) and c) follow from perturbation theory. We see that Theorem 2 provides the necessary input to Theorem 1. So we can conclude

Theorem 3: a) $h_{\underline{\nu}}(p;m_o,\lambda)$ is C^∞ in m_o,λ and analytic in p in the axiomatic domain. The derivatives are given (for ∂_λ and $P(\phi) = \phi^4$) by

$$\partial_\lambda h_{\underline{\nu}}(p_1,\dots p_n;m_o\lambda) = -h_{\underline{\nu}\cup 4}(p_1,\dots p_n,o;m_o,\lambda).$$

b) The physical mass $m(m_o,\lambda)$ and the field strength $Z(m_o,\lambda)$ are C^∞ in m_o and λ.

c) The function $m(m_o,\lambda)$ has an inverse function $m_o(m,\lambda)$ in the sense that

$$m(m_o(a,\lambda)\ ,\ \lambda) = a\ ,$$

and $m_o(a,\lambda)$ is C^∞ in λ.

d) $\prod_{j=1}^{n} (p_j^2 - m^2)\, h_{1\dots 1}(p;\ m_o(m,\lambda),\lambda)$ is C^∞ in λ and holomorphic in $p_1,\dots p_n$ in the "axiomatic domain" of the n-point function with thresholds above $\sqrt{2}m$ and no single particle poles (at $p_j^2 = m^2$). In particular the derivatives have no single particle poles.

e) The expression (2) for the S-matrix elements is C^∞ in λ for fixed physical mass m, whenever it exists, as a restriction of the C^∞ function $Z^{-n}(m_o(m,\lambda),\lambda) \prod_{j=1}^{n} (p_j^2-m^2)\, h_{1\dots 1}(p,m_o(m,\lambda),\lambda)$ (to the mass shell) (i.e.e.g. for non overlapping momenta (Hepp)).

Proofs: a) Follows by Theorem 1b), (The derivative w.r.t. λ commutes with Fourier transform) and the uniformity of the bounds in λ (Thm 2c).

Note: This crucially depends on the fact that the derivative of $\overset{\rho T}{S}$ involves the "correct" extension

S^T and not another one!

b) Follows by contour integrals of $h_{1,1}$ (parts of this result have been shown earlier by Spencer).

c) By scaling, for some f(t) which is C^∞,

$$\partial_{m_o^2} m^2(m_o,\lambda) = f(\lambda/m_o^2) - (\lambda/m_o^2) f'(\lambda/m_o^2) \neq o \text{ if } \lambda/m_o^2$$

is small. c) now follows from b) and the implicit function theorem.

d) This follows from the principle: if a function $f(z,\lambda)$ is holomorphic in $|z| < 1$ and bounded in $\lambda \varepsilon [o,1]$, and for $|z| > \frac{1}{2}$ differentiable in λ with derivative holomorphic in $|z| > \frac{1}{2}$ then the same is true for <u>all</u> $|z| < 1$.

e) Follows from d), and the fact that restrictions of differentiable functions (distributions) are differentiable.

<u>Corollary 4</u>: <u>Ordinary perturbation theory is asymptotic to the S-matrix elements at</u> $\lambda = o$.
This follows from the fact that the same is true for Schwinger functions, and that perturbation theory has the axiomatic analyticity properties.

We see thus that the S-matrix is nontrivial if perturbation theory is nontrivial.

References

GJS1 J. GLIMM, A. JAFFE, T. SPENCER: The particle structure of the weakly coupled $P(\phi)_2$ model and other applications of high temperature expansions, in: Constructive Quantum Field Theory, G. Velo, A. Wightman, eds., Springer Lecture Notes in Physics, Vol. 25, Berlin-Heidelberg-New-York 1973.

GJS2 J. GLIMM, A.JAFFE, T. SPENCER: The Wightman axioms and particle structure in the $P(\phi)_2$ quantum field model; Ann. Math. 100, 585 (1974).

S T. SPENCER: The decay of the Bethe-Salpeter kernel in $P(\phi)_2$ quantum field models. Harvard 1975.

G V. GLASER: On the equivalence of the Euclidean and Wightman formulation of field theory. Commun. math. Phys. 37, 257 (1974).

OS K. OSTERWALDER, R. SCHRADER: Axioms for Euclidean Green's functions II. Commun. math. Phys. 42, 281 (1975).

D J. DIMOCK: Asymptotic perturbation expansion in the $P(\phi)_2$ quantum field theory: Commun. math. Phys. 35, 347 (1974).

EG H. EPSTEIN, V. GLASER: Le rôle de la localité dans la renormalisation perturbative en théorie quantique des champs, in Statistical Mechanics and Quantum Field Theory, C. de Witt and R. Stora Eds., Gordon and Breach, 1971. R. Stora Eds., The role of locality in perturbation theory Ann. Inst. Henri Poincaré 19, 211 (1973).

BEG J. BROS, H. EPSTEIN, V. GLASER: Local analycity properties of the n-particle scattering amplitude. Helv. Phys. Acta 45, 149 (1972).

H K. HEPP: On the connection between the LSZ and Wightman quantum field theory. Commun. Math. Phys. 1, 95 (1965).

R D. RUELLE: Connection between Wightman functions and Green Functions in p-space. Il Nuovo Cimento, 19, 356 (1961).

BEGS J. BROS, H. EPSTEIN, V. GLASER, R. STORA, n-point functions in local quantum field theory, to appear in proceedings Les Houches Summer Institute, D. Iagolnitzer ed. (1975).

Acta Physica Austriaca, Suppl. XVI, 69-86 (1976)

Time Ordered Operator Products and the Scattering Matrix in $P(\phi)_2$ Models

Konrad Osterwalder[+]
Jefferson Laboratory of Physics
Harvard University
Cambridge, Mass., U.S.A.

[+] Work supported by the National Science Foundation under grant MPS 73-05037
Alfred P. Sloan Foundation Fellow

I. Introduction

It has been known for more than two years now that weakly coupled $P(\phi)_2$ models have an isolated one particle hyperboloid and hence by the Haag-Ruelle theory they possess a well defined scattering matrix.

The main purpose of this talk is to present a simple and pedestrian method of proving that this S-matrix is non trivial. This method has been worked out in collaboration with Roland Sénéor [8], see also the contributions of Eckmann and of Dimock to these proceedings. We proceed in three steps.

1) Starting from the Euclidean Green's functions $\mathfrak{S}_n(\underset{\sim}{x})$ we construct the Wightman functions $\mathcal{W}_n(\underset{\sim}{z})$ which are analytic in the standard (axiomatic) domain. Then the time ordered Green's functions $\tau_n(\underset{\sim}{x})$ can be defined as boundary values of the Wightman functions. Our procedure can easily be generalized to show the existence of time ordered operator products $T(\phi(x_1)\ldots\ldots\phi(x_n))$ as densely defined operator valued distributions.
2) In a second step we expand the (truncated) Euclidean Green's functions such that one-particle contributions from external lines are exhibited explicitly. We write

$$\mathfrak{S}_n(x_1\ldots x_n)^T =$$

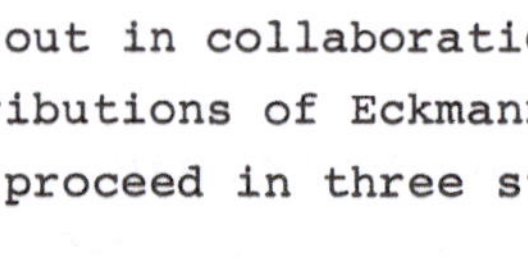

$$= \int \prod_{i=1}^{n} dy_i \; C(x_i - y_i) \sum_{k=1}^{n} \lambda^k \, F_k(\underset{\sim}{y},\lambda) \qquad (1)$$

where $C(x-y) = (-\Delta+m^2)^{-1}(x,y)$, λ is the coupling constant and is restricted to an interval $[0,\lambda_o]$ (weak coupling) and the functions F_k will obey bounds which are <u>uniform</u> in λ.

3) In the third step we analytically continue (1) to real times. This gives us the truncated time ordered Green's functions τ_n on the l.h.s. of the equation and Feynman propagators Δ_F replacing the factors C on the right, while the coefficient functions F_k go over into generalized time ordered Green's functions $\hat{F}_k$. Again we have uniform bounds on $\hat{F}_k$. Amputation of this τ_n then just cancels out all the propagators Δ_F. Finally the amputated τ_n have to be restricted to the mass shell. This we can do for nonoverlapping momenta using the method of Hepp [5]. As a result we can write the connected part of the S-matrix element $<p_1^{in}\ldots p_n^{in}|p_1^{out}\ldots p_m^{out}>$ in the form $\sum_{k=1}^{n+m} \lambda^k \hat{F}_k$ with all the $\hat{F}_k$ bounded uniformly for $\lambda \in [o,\lambda_o] \cdot \hat{F}_1$ is just the first term in the ordinary power series expansion of the S-matrix element. This shows that $S \neq \mathbb{1}$.

II. Time Ordered Green's Functions and Time Ordered Operator Products.

Euclidean Green's functions for a $\lambda P(\phi)_2$ model are defined by

$$\mathfrak{G}_n(\underset{\sim}{x}) = \langle \prod_{i=1}^{n} \Phi(x_i) \rangle = \lim_{\Lambda \to \mathbb{R}^2} \frac{\int \prod_{i=1}^{n} \Phi(x_i)\, e^{-\lambda \int_\Lambda :P(\Phi):(x)\, d^2x}\, d\mu_o(\Phi)}{\int e^{-\lambda \int_\Lambda :P(\Phi):(x)\, d^2x}\, d\mu_o(\Phi)} \tag{2}$$

where $\underset{\sim}{x} = (x_1, \ldots x_n)$, $x_i = (x_i^o, \vec{x}_i) \in \mathbb{R}^2$. μ_o is the Gaussian measure on $\mathcal{S}'(\mathbb{R}^2)$ with mean 0 and covariance $C = (-\Delta + m_o^2)^{-1}$. Glimm, Jaffe and Spencer have shown [3,4] that for $m_o > 0$ fixed there exists a $\lambda_o > 0$ such that for all $\lambda \in [0, \lambda_o]$ (2) exists, satisfies all the axioms for Euclidean Green's functions [7] and defines a quantum field theory whose mass operator $M = (H^2 - \vec{P}^2)^{1/2}$ nas eigenvalues 0, m and no other spectrum in $[0, m']$. Here $m' = 2m_o - \varepsilon$. and $m \in (m_o - \varepsilon, m_o + \varepsilon)$ for some $\varepsilon = \varepsilon(\lambda_o)$ which tends to zerc as λ_o gets small. It is easy to see that without loss of generality we may always assume that $m = m_o$.

The axioms for Euclidean Green's functions (in particular: positivity, Euclidean covariance and regularity) allow us to analytically continue the $\mathfrak{G}_n(\underline{x})$ in all the "time variables" x_i^o to all of $\mathbb{C}^n$ minus points where $\mathrm{Re}\, x_i^o = \mathrm{Re}\, x_j^o$ for some $i \neq j$. The reconstruction theorem [7] tells us that the boundary value of these analytic functions are the Wightman distributions of the theory. More precisely

$$\mathcal{W}_n(\underset{\sim}{x}) = \lim_{\substack{y_1^o<y_2^o<\ldots<y_n^o \\ |y_i|\to o}} \mathfrak{S}_n(y_1^o+ix_1^o,\vec{x}_1,\ldots.y_n^o+ix_n^o,\ \vec{x}_n \qquad (3)$$

in the sense of distributions. The direction of approach to the boundary is crucial. If we chose a different direction we get something else. E.g. we claim that we obtain the <u>time ordered Green's</u> functions as follows.

$$\left\langle T\prod_{i=1}^{n}\phi(x_i)\right\rangle = \lim_{\mu\to i}\ \mathfrak{S}_n\ (\mu x_1^o,\vec{x}_1,\ldots\mu x_n^o,\vec{x}_n) \qquad (4)$$

Here $\mu\to i$ means $\mu = i-\delta\ ,\delta\not\downarrow 0$; $\phi(x)$ is the relativistic field operator and T means time ordering. Formally the l.h.s. of (4) is defined by

$$\left\langle T\prod_{i=1}^{n}\phi(x_i)\right\rangle = \sum_{\pi}\prod_{i=1}^{n-1}\Theta(x^o_{\pi(i)}-x^o_{\pi(i+1)})\left\langle\prod\phi(x_{\pi(i)})\right\rangle \quad (5)$$

where the sum $\sum_{\pi}$ runs over all permutation π of $(1,2...n)$, $\Theta(.)$ is the Heaviside step function and $\left\langle\prod_i\phi(x_i)\right\rangle = \mathcal{W}_n(\underset{\sim}{x})$. To see the (formal) consistency of eqs. (2), (3) and (4) we consider the sector where $x^o_{\pi(1)} > x^o_{\pi(2)}\ldots x^o_{\pi(n)}$ for some π. There

$$\mathrm{Re}\mu x^o_{\pi(i)} = -\delta x^o_{\pi(i)} < -\delta x^o_{\pi(i+1)} = \mathrm{Re}\mu x^o_{\pi(i+1)}.$$

Hence by (4)

$$\left\langle T\prod_{i=1}^{n}\phi(x_i)\right\rangle = \lim_{\substack{\mathrm{Re}Z^o_{\pi(1)}<\ \mathrm{Re}Z^o_{\pi(2)}<\ldots\mathrm{Re}Z^o_{\pi(n)} \\ z_i^o = y_i^o+ix_1^o \\ y_i^o\to 0}} \mathfrak{S}_n(z_1^o,\vec{x}_1,\ldots.z_n^o,\vec{x}_n)$$

$$= \mathcal{W}_n(x_{\pi(1)}\ldots.x_{\pi(n)}) \quad , \qquad \text{by (3)}$$

This agrees with (5).

While the right hand side of (5) is a priori ill defined (multiplication of a distribution with a step function) it turns out that we can use (4) to define time ordered Green's functions. (Alternatively we could show that (5) is meaningful)

Theorem 1:

Eq. (4) defines time ordered Green's function as tempered distributions with bounds uniform in λ for $\lambda \in [0,\lambda_o]$.

Proof:

We construct the analytic continuation $\mathfrak{G}_n(\underset{\sim}{z})$ $(z_i = z_i^o, \vec{x}_i)$ of $\mathfrak{G}_n(x)$ in the standard inductive manner, see [7].

Estimates in [3] and [7] show that

$$|\mathfrak{G}_n(\underset{\sim}{x})| \leq \alpha^n \; n! \; (1-\ln \; \varepsilon)^{2n}$$

for some constant α not depending on λ for $\lambda \in [0,\lambda_o]$ and

$$\varepsilon = \min \{1, \; |x_i^o - x_j^o| \quad \text{for } 1 \leq i < j \leq k\}$$

To get a bound on the analytic continuation we look at a sector in $\mathbb{R}^{2k}$ where

$$x_1^o < x_2^o \ldots < x_n^o$$

(Other sectors are dealt with similarly). In this sector, by translation invariance,

$$\mathfrak{G}_n(\underset{\sim}{x}) = S_{n-1}(\underset{\sim}{\xi}) \quad , \quad \underset{\sim}{\xi} = (\xi_1, \ldots \xi_{n-1})$$
$$\xi_i = x_{i+1} - x_i$$

The analytic continuation will be

$$\mathfrak{G}_n(\underset{\sim}{z}) = S_{n-1}(\underset{\sim}{\zeta}) \text{ with } \mathrm{Re}\; \zeta_i^o = \xi_i^o > 0.$$
$$\vec{\zeta}_i = \vec{\xi}_i \text{ real}$$

According to [7] the analytic continuation of $S_{n-1}(\underset{\sim}{\xi})$ is obtained in an infinite sequence of steps. After the N'th step one has defined $S_{n-1}(\underset{\sim}{\zeta})$ in a domain of analyticity $(\zeta_1^o,\ldots\zeta_{n-1}^o) \in C_{n-1}^{(N)}$. The domains $C_{n-1}^{(N)}$ strictly increase in N and eventually fill all of $\mathbb{C}_+^{n-1} = \{z:\text{Re}z>0\}^{n-1}$. Also we claim that in $C_{n-1}^{(N)}$

$$|S_{n-1}(\underset{\sim}{\zeta})| \leq \alpha^n\, n!\, 2^{nN}\, (1-\ln\varepsilon')^{2n} \tag{6}$$

where

$$\varepsilon' = \min\,\{1, |\text{Re}\zeta_i^o|,\ i=1,\ldots n-1\}.$$

We now explain to prove (6) and how to control the factor 2^{nN} which seems to get out of control as $N\to\infty$. The basic idea is to use the fact that the functions S_{n-1} may be interpreted as scalar products of vectors in the physical Hilbert space $\mathcal{H}$.

<u>Lemma:</u>

There are $\mathcal{H}$-vector valued functions $\Psi_k(x,\zeta_1,\ldots\zeta_{k-1})$ such that

$$(\Psi_\ell(x,\zeta_1,,,\zeta_{\ell-1}),\ \Psi_{n-\ell}(x',\zeta_{\ell+1}\ldots\zeta_{n-1})) =$$

$$= S_{n-1}(\hat\zeta_{\ell-1}\ldots\hat\zeta_1,\ x+x';\zeta_{\ell+1}\ldots\zeta_{n-1})$$

Here $\hat\zeta_i = (\bar\zeta_i^o,\vec\zeta_i)$, Re $\zeta_i^o > o$, $\vec\zeta_i = \xi_i$ real. (7)

Now suppose that we have for all n constructed $S_{n-1}(\underset{\sim}{\zeta})$ for $\underset{\sim}{\zeta} \in C_{n-1}^{(N)}$ and that we have verified ineq. (6). By the lemma,

$$S_{n-1}(\zeta_1,\ldots\zeta_{\ell-1},\ 2x^o+iy^o,\ 2\vec x,\ \zeta_{\ell+1},\ldots\zeta_{n-1})$$

$$\equiv (\Psi_\ell(x,\hat\zeta_{\ell-1}\ldots\zeta_1),\ e^{-iy^oH}\ \Psi_{n-\ell}(x,\zeta_{\ell+1},\ldots\zeta_{n-1}))$$

defines an analytic continuation of S_{n-1} in the variable ξ_ℓ^o to $\zeta_\ell^o \in \mathbb{C}_+ = \{z|\text{Re } z > 0\}$. A bound on this continuation follows from the Schwarz inequality

$$|S_{n-1}(\ldots,2x^o+iy,\ldots)| \le ||\Psi_\ell(\ldots)||\ ||\Psi_{n-\ell}(\ldots)||$$

$$= [S_{2\ell-1}(\zeta_1,\ldots\zeta_{\ell-1},2x,\hat{\zeta}_{\ell-1}\ldots\hat{\zeta}_1)\,.$$

$$\times\ S_{2(n-\ell)-1}(\hat{\zeta}_{n-1}\ldots\hat{\zeta}_{\ell+1},\ 2x,\zeta_{\ell+1}\ldots\zeta_{n-1})]^{1/2}$$

(by the lemma)

$$\le \alpha^n\ 2^{nN}(1-\ln\varepsilon')^{2n}\ [(2\ell)!(2(n-\ell))!]^{1/2}$$

by the inductive assumption (6)

$$\le \alpha^n\ 2^{nN}(1-\ln\varepsilon')^{2n}\ (n!2^n)$$

$$= \alpha^n\ 2^{n(N+1)}n!\ (1-\ln\varepsilon')^{2n} \tag{8}$$

Inequality (8) is the same as (6) with N+1 replacing N. Repeating the analytic continuation procedure for all variables $\zeta_1^o,\ldots\zeta_{n-1}^o$ and then taking the domain of holomorphy gives the new bigger domain of analyticity $C_{n-1}^{(N+1)}$ with the corresponding bound (8). This bound is not changed if we pass to the domain of holomorphy, because of the maximum principle, see [7]. To eliminate the factor 2^{nN} from the estimate (6) we have to use the dependence of the size of $C_k^{(N)}$ on N. A complicated geometrical argument which I don't want to reproduce here (see however [7]) shows that for large N:

$C_k^{(N)}$ contains the points with

$$\underset{\sim}{\zeta}\ :\ |\arg\zeta_i^o| < \frac{\pi}{2}\ (1-2^{-N/2}\ \gamma_k),\ i=1,2\ldots k$$

for some constant $\gamma_k \in [0,1]$.

Given a point $\underset{\sim}{\zeta}$ we set

$$\frac{\pi}{2}\ (1-\rho)\ =\ \max\ |\arg\ \zeta_i^o| \tag{9}$$

and choose N such that

$$\frac{\pi}{2}\,(1-2^{-(N-1)/2}\,\gamma_k) \leq \frac{\pi}{2}\,(1-\rho) \leq \frac{\pi}{2}\,(1-2^{-N/2}\,\gamma_k).$$

Then $\underset{\sim}{\zeta} \in C_k^{(N)}$ and $\rho \leq \gamma_k\, 2^{-(N-1)/2}$, i.e.

$$2^{kN} \leq \left(\frac{\gamma_k\sqrt{2}}{\rho}\right)^{2k} \tag{10}$$

We finally want an estimate on

$$\mathfrak{G}_n(\mu x_1^o,\, \vec{x}_1, \ldots \mu x_n^o, \vec{x}_n) =$$

$$S_{n-1}(\mu\xi_1^o, \vec{\xi}_1, \ldots \mu\xi_{n-1}^o, \vec{\xi}_{n-1})$$

$\mu = i-\delta$, $x^o_{\pi(1)} > x^o_{\pi(2)} \ldots > x^o_{\pi(n)}$ for some permutation π. We set $\xi_i = x_{\pi(i+1)} - x_{\pi(i)}$, then Re $\mu\xi_i^o = \delta(x_{\pi(i)} - x_{\pi(i+1)}) > 0$, as required.

Furthermore

$$|\arg\, \mu\xi_i^o| = |\arg(-\mu)|$$

and from (9), with $\mu = i-\delta$, $0 < \delta << 1$,

$$\rho = 1 - \frac{2}{\pi}\,|\arg(-\mu)| = 1 - \frac{2}{\pi}\,\text{arctg}\,\delta \sim \frac{2}{\pi}\,\delta$$

Hence from (10)

$$2^{kN} \leq \left(\gamma_k\sqrt{2}\cdot\frac{\pi}{2\delta}\right)^{2k} = c_k\,\delta^{-2k} \tag{11}$$

Furthermore

$$\varepsilon' = \min\,\{1,\, |\text{Re}\mu\xi_i^o|\} \geq \delta\,\min\,\{1, |\xi_i^o|\} = \delta\cdot\varepsilon$$

and

$$1-\ln\varepsilon' \leq 1-\ln\delta\cdot\varepsilon \leq (1-\ln\varepsilon)(1-\ln\delta)\cdot \tag{12}$$

Substituting (11) and (12) into (6) we obtain

$$|\mathfrak{S}_n(\mu x_1^o,\vec{x}_1,\ldots.\mu x_n^o,\vec{x}_n)| \le \alpha^n \; n! \; c_n \delta^{-2n} \; |(1-\ln\varepsilon)(1-\ln\delta)|^{2n}$$

$$\le c_n' \; (1-\ln\varepsilon)^{2n} \; \delta^{-2n-1/2} \qquad (13)$$

for some new constant c_n', $\varepsilon = \min\{1; |x_i^o - x_j^o|, \; i \neq j\}$ and $\mu = i-\delta$. This bound holds for all values of $\underset{\sim}{x}$ as long as $\varepsilon \neq o$. Now it is easy to show that the limit (4) exists and defines a tempered distribution. For $f \in \mathcal{S}(\mathbb{R}^{2n})$ and $\mu = i-\delta$ set

$$\tau_\delta(f) = \int \mathfrak{S}_n(\mu x_1^o,\vec{x}_1,\ldots.\mu x_n^o,\vec{x}_n) f(\underset{\sim}{x}) \; d\underset{\sim}{x}$$

(Notice that the local singularities of $\mathfrak{S}_n$ are integrable). Then

$$\frac{d}{d\delta}\tau_\delta(f) = \int \frac{d}{d\delta} \; \mathfrak{S}_n((i-\delta)x_1^o,\ldots.) \; f(\underset{\sim}{x}) \; d\,\underset{\sim}{x}$$

$$= \int \mathfrak{S}_n(\mu x_1^o,\ldots) \sum_i \left[\frac{\partial}{\partial x_i^o} \; \left(\frac{x_i^o}{\mu}\right) \; f(\underset{\sim}{x})\right] d\underset{\sim}{x}$$

after (2n+1)-fold differentiation we obtain

$$\frac{d^{2n+1}}{d\delta^{2n+1}}\tau_\delta(f) = \int \mathfrak{S}_n(\,x_1^o,\ldots)\mu^{-(2n+1)} \sum_{k=1}^{2n+1} c_k \; \times$$

$$\sum_{i_1,\ldots i_k} \left(\prod_{\gamma=1}^{k} \frac{\partial}{\partial x_{i_\gamma}^o} \; x_{i_\gamma}^o\right) f(\underset{\sim}{x}) \; d\underset{\sim}{x} \qquad (14)$$

where c_k are combinatorial factors. Using ineq. (13) we conclude that

$$\left|\frac{d^{2n+1}}{d\delta^{2n+1}}\tau_\delta(f)\right| \le \text{const.} \; \delta^{-(2n+1/2)} \; |f|$$

for some Schwartz norm $|.|$. Integrating (14) (2n+1) times we find that $\tau_\delta(f)$ is uniformly bounded as $\delta \downarrow 0$ and converges to a tempered distribution. This concludes our proof of theorem 1.

Theorem 1 can easily be generalized to establish the existence of vacuum expectation values of arbitrary products of time ordered operator products. For x a k-tuple of vectors in $\mathcal{R}^2$ we formally set

$$T\phi(x) = \sum_{\pi} \prod_{i=1}^{k-1} \theta(x^o_{\pi(i)} - x^o_{\pi(i+1)}) \; \phi(x_{\pi(1)}) \ldots \phi(x_{\pi(k)}) \tag{15}$$

To define $T\phi(x)$ as an operator with dense domain $\mathcal{D}_T$ it suffices to define all the vacuum expectation values of products of time ordered operator products and to establish an appropriate positivity condition. Such a definition is readily available in terms of boundary values of analytically continued Euclidean Green's functions. As in the proof of Theorem 1 we show that

$$\langle T\phi(x^{(1)}) \cdot T\phi(x^{(2)}) \ldots\ldots T\phi(x^{(n)}) \rangle$$

$$\overset{\text{def.}}{=} \lim_{\delta \downarrow 0} \mathfrak{G}(\mu \underset{\sim}{x}^{(1)} + \underset{\sim}{\delta}, \; \mu \underset{\sim}{x}^{(2)} + 2\underset{\sim}{\delta}, \ldots \; \mu \underset{\sim}{x}^{(n)} + n\underset{\sim}{\delta}) \tag{16}$$

exists as a tempered distribution. Here $\underset{\sim}{x}^{(j)}$ is a k_j-tuple $(x_1^{(j)}, \ldots x_{k_j}^{(j)})$ of vectors in $\mathcal{R}^2$ and $\mu \underset{\sim}{x}^{(j)} + j\underset{\sim}{\delta}$ stands for the quantity

$$(\mu x_1^{(j)o} + j\delta, \; \vec{x}_1^{(j)}, \ldots\ldots \mu x_{k_j}^{(j)o} + j\delta, \; \vec{x}_{k_j}^{(j)});$$

μ is now chosen to be $\mu = i - \delta^2$. We leave it as an exercise to show that

1) the positivity properties of the Euclidean Green's functions and of their analytic continuations imply that the right hand side of (16) does indeed define the vacuum expectation value of a product of operators $T\phi(\underset{\sim}{x}^{(j)})$.
2) by (3), the operators $T\phi(\underset{\sim}{x})$ as defined by (16) agree with the formal definition (15) whenever smeared with a test function f that vanishes with all its derivations at points of equal times $(x_i^o = x_j^o, \; i \neq j)$
3) the domain $\mathcal{D}_T$ spanned by vectors of the form

$$\prod_j \left(\int T\phi(\underset{\sim}{x}^{(j)}) \; f_j(\underset{\sim}{x}^{(j)}) \; d\underset{\sim}{x}^{(j)} \right) \Omega, \; f_j \in \mathcal{S},$$

is a dense, invariant domain for smeared time ordered operator products.

4) all the expected properties such as Lorentz covariance and locality hold

We summarize some of these results in

Theorem 2:

The limit (16) exists in the sense of distributions and defines the time ordered operator products $T\phi(\underset{\sim}{x})$ as operator valued distributions with invariant domain $\mathcal{D}_T$.

III. Expanding Euclidean Green's Functions and S-Matrix Elements

In this section we derive an expansion for the Euclidean Green's functions that will enable us to exhibit the amputation of the time ordered Green's functions explicitly. The expansion could be iterated; it would then generate the standard Feynman perturbation expansion.

Let us introduce the notations

$$V(x) \equiv :P(\Phi(x)):$$

$$V^{\ell}(x) \equiv :P^{(\ell)}(\Phi(x)): \quad \text{where } P^{(\ell)}(\xi) \equiv \frac{d^{\ell}}{d\xi^{\ell}} P(\xi)$$

and P is the polynomial with even leading coefficient, introduced in the definition (2) of the Euclidean Green's functions. The basic tool of our expansion is the "integration by parts" formula [2]

$$\left\langle \Phi(x_1) \prod_{i=2}^{n} \Phi(x_i) \right\rangle = \int C(x_1-y_1) \left\langle \frac{\delta}{\delta\Phi(y_1)} \prod_{i=2}^{n} \Phi(x_i) \right\rangle$$

$$- \lambda \int C(x_1\ y_1) \left\langle V^1(y_1) \prod_{i=2}^{n} \Phi(x_i) \right\rangle \tag{17}$$

Remember that C is the kernel of $(-\Delta+m^2)^{-1}$, where m is the physical mass.

After repeated application of formula (17) (for all the factors $\Phi(x_i)$) and considering only the truncated part we obtain

$$\left\langle \prod_{i=1}^{n} \Phi(x_i) \right\rangle^{T} = \sum_{\pi} (-\lambda)^{|\pi|} \int \prod_{\sigma\in\pi} \left[dy_\sigma \prod_{i\in\sigma} C(x_i-y_\sigma)\right]$$

$$\left\langle \prod_{\sigma\in\pi} V^{|\sigma|}(y_\sigma) \right\rangle \tag{18}$$

Here the sum runs over all partitions π of $\{1,2,\ldots n\}$ into $|\pi|$ nonempty mutually disjoint sets σ, with $|\pi|$ taking all values $1,2,\ldots n$; $|\sigma|$ is the number of elements in σ; $\langle\ldots\rangle^T$ means truncated vacuum expectation value. As the $V^{|\sigma|}(y_\sigma)$ are themselves products of field operators, we could go on using (17), thus generating the ordinary perturbation expansion.

Next we want to "analytically continue" equation (18). We know how to continue each of the factors on the r.h.s. of (18) separately. With $\omega(\vec{p}) = \sqrt{\vec{p}^2+m^2}$, $px = p^o x^o - \vec{p}\vec{x}$,

$$C_\mu(x) \equiv C(\mu x^o, \vec{x}) = \frac{\mu}{2\pi} \int \frac{e^{ipx}}{p_o^2+\mu^2\omega^2}\, d^2p \tag{19}$$

is the analytic continuation of C. The truncated vacuum expectations $\left\langle \prod_{\sigma\in\pi} V^{|\sigma|}(y_\sigma)\right\rangle^T$ will be denoted by $\mathfrak{G}_\pi^T(\underset{\sim}{y})$ and their analytic continuation to complex times z_σ^o will be denoted by $\mathfrak{G}_\pi^T(z)$. It is easy to see that the local singularities of $\mathfrak{G}_\pi^T(z)$ are at most logarithmic. Hence the function

$$\sum_\pi (-\lambda\mu)^{|\pi|} \int_{G_\varepsilon} \prod_{\sigma\in\pi} [dy_\sigma \prod_{i\in\sigma} C_\mu(x_i-y_\sigma)]\ \mathfrak{G}_\pi^T(\mu\underset{\sim}{y}) \tag{20}$$

is analytic in $\mu\in M=\{\mu : \mathrm{Re}\,\mu<0,\ 1/2<|\mu|<3/2\}$ and uniformly bounded in ε for μ in any compact subset of M, $\underset{\sim}{x}$ fixed and

$$G_\varepsilon = \{\underset{\sim}{y}\ |\ |y_\sigma^o-y_{\sigma'}^o| > \varepsilon,\ |y_\sigma^o-x_i^o| > \varepsilon \text{ for all } \sigma\neq\sigma',\ i\in\sigma\}$$

By Vitali's theorem we may therefore take the limit $\varepsilon\downarrow 0$ in (20) and obtain an analytic function of μ, which for real μ equals (18) with all the x^o and y^o-variables scaled by μ. By the uniqueness of analytic continuation we find that for $\mu\in M$

$$\mathfrak{G}_n^T(\mu\underset{\sim}{x}) = \sum_\pi (-\lambda\mu)^{|\pi|} \int \prod_{\sigma\in\pi} [dy_\sigma \prod_{i\in\sigma} C_\mu(x_i-y_\sigma)]\,\mathfrak{G}_\pi^T(\mu\underset{\sim}{y}). \tag{21}$$

In the limit $\mu\to i$; the l.h.s. of (21) leads to the truncated time ordered Green's functions $\tau_n(\underset{\sim}{x})^T$ as discussed before. The factors $C_\mu(x)$ just become the Feynman propagators $\Delta_F(x)$ = Fourier transform of $(p^2-m^2+i\varepsilon)^{-1}$. The "generalized" Euclidean Green's functions $\mathfrak{G}_\pi^T(\mu y)$ finally lead to generalized time ordered products

$$\left\langle T \prod_{\sigma\in\pi} V^{|\sigma|}(y_\sigma)\right\rangle^T, \text{ where } V^{\ell}(y) = :P^{(\ell)}(\phi(y)):$$

($\phi(y)$ is the relativistic field operator).

According to the LSZ reduction formulas [6,5] the scattering matrix elements are the amputated time ordered Green's functions with all the energy-momentum variables restricted to the mass shell. With the expansion (21), $\mu\to i$, in mind we see that amputation just eliminates all the factors $\Delta_F = \lim_{\mu\to i} C_\mu$. Then it remains to show that the Fourier transform of $\left\langle T \prod V^{|\sigma|}(y_\sigma)\right\rangle^T$ can be restricted to the mass shell. In [5] Hepp has shown how to do this rigorously for the case where in-going and out-going particles have non-overlapping velocities. Here we only sketch the argument, for details see [8]. We let $\tilde{\mathcal{S}}_m$ be the space of test functions $\tilde{f}\in\mathcal{S}(\mathbb{R}^2)$ with supp $\tilde{f}\subset\{p^0>0 \,|\, 0<p^2<m'^2\}$. (Remember that $m'\in(m,2m)$ was chosen such that the mass operator M has no spectrum in the interval $[0,m')$ except for the eigenvalues 0 and m.) For $\tilde{f}\in\tilde{\mathcal{S}}_m$ $\omega=\sqrt{\vec{p}^2+m^2}$, we define

$$f(x,t) = (2\pi)^{-1}\int e^{ipx}\tilde{f}(p)e^{i(p^0-\omega)t}\,d^2p$$

$$\hat{f}(\vec{p}) = \tilde{f}(\omega,\vec{p}); \quad \dot{f}(x,t) = \frac{\partial f}{\partial t}(x,t)$$

We call a sequence $\{\tilde{f}_i\}\in\tilde{\mathcal{S}}_m$ non-overlapping if for all $p_i\in$ supp $\tilde{f}_i$, $\omega_i = \omega(\vec{p}_i)$,

$$\frac{\vec{p}_i}{\omega_i} \neq \frac{\vec{p}_j}{\omega_j} \quad \text{for all } i \neq j.$$

By Hepp's analysis, for non-overlapping $\{\tilde{f}_i\}$, the connected part of the S-matrix element

$$\left\langle \hat{f}_1^{out}, \dots \hat{f}_n^{out} \,\middle|\, \hat{f}_{n+1}^{in} \dots \hat{f}_{n+m}^{in}\right\rangle \tag{22}$$

is given by

$$\text{const.} \int \prod_{i=1}^{n+m} dt_i \int \prod_{i=1}^{n} [dx_i \, \dot{\bar{f}}_i(x_i,t_i)] \int \prod_{i=n+1}^{n+m} [dx_i \, \dot{f}_i(x_i,t_i)]$$

$$\times \left\langle T \prod_{i=1}^{n+m} \Phi(x_i) \right\rangle^T \tag{23}$$

Now we may substitute the expansion (21) with $\mu \to i$ into (23) to obtain an expansion of the scattering matrix elements. We introduce

$$\tilde{g}_j(p) = - \frac{\hat{f}_i(p)}{p^o + \omega(\vec{p})}$$

and

$$g_j(x,t) = (2\pi)^{-1} \int e^{ipx} \, \tilde{g}_j(p) e^{i(p^o - \omega)t} \, d^2p$$

clearly $\{g_j\}$ is non-overlapping if $\{\tilde{f}_j\}$ is, and

$$\dot{f}_j = \lim_{\mu \to i} g_j * C_\mu^{-1} \tag{24}$$

in the topology of $\mathcal{S}$.

Substitution of (24) and of (21) into (23) gives

$$\text{const.} \int \prod_{i=1}^{n+m} dt_i \{ \sum_\pi (-i\lambda)^{|\pi|} \int \prod_{\sigma \in \pi} dy_\sigma \prod_{i=1}^{n} (-\bar{g}_i)(y_i,t_i)$$

$$\prod_{i=n+1}^{n+m} g_i(y_i,t_i) \left\langle T \prod_{\sigma \in \pi} V^{|\sigma|}(y_\sigma) \right\rangle^T \} \tag{23'}$$

We claim that the t_i-integrations and the sum $\sum_\pi$ can be interchanged, for a proof we refer to [8]. The following theorem is an immediate consequence.

<u>Theorem 3:</u>

For non-overlapping $\{\tilde{f}_i\} \subset \mathcal{S}_m$ the connected part of the S-matrix element

$$\left\langle \hat{f}_1^{out}, \dots \hat{f}_n^{out} \mid \hat{f}_{n+1}^{in}, \dots \hat{f}_{n+m}^{in} \right\rangle$$

is given by

$$\text{const.}\sum_{\pi}(-i\lambda)^{|\pi|}\int\prod_{i=1}^{n+m} dt_i \ \{ \int\prod_{\sigma\in\pi} dy_\sigma \prod_{i=1}^{n}(-\bar{g}_i)(y_i,t_i)$$

$$\prod_{i=n+1}^{n+m} g_i(y_i,t_i) \quad \left\langle T \prod_{\sigma\in\pi} V^{|\sigma|}(y_\sigma)\right\rangle^T \} \tag{25}$$

The coefficients of $\lambda^{|\pi|}$ in (25) are bounded uniformly in λ for $\lambda\in[0,\lambda_0]$.

Theorem 3 implies that for an interaction polynomial $P(\xi) = \xi^{2k}$ + lower order terms and for non-overlapping $\{\hat{f}_i\}$,

$$\left\langle \hat{f}_1^{out}\ldots\hat{f}_m^{out} \mid \hat{f}_{m+1}^{in}\ldots\hat{f}_{2n}^{in}\right\rangle \text{ connected}$$

$$= -i\lambda(2k)!\int\prod_{i=1}^{m}\frac{\hat{f}_i(\vec{p}_i)}{2\omega_i}\, d\vec{p}_i \prod_{i=m+1}^{2n}\frac{\hat{f}_i(\vec{p}_i)}{2\omega_i}$$

$$\times\ \delta\Big(\sum_{i=1}^{m}\omega_i - \sum_{m+1}^{2n}\omega_i\Big)\ \delta\Big(\sum_{i=1}^{m}\vec{p}_i - \sum_{m+1}^{2n}\vec{p}_i\Big)$$

$$+\ O(\lambda^2)$$

Hence for λ sufficiently small, the S-matrix is non-trivial.

References

[1] J. GLIMM, A. JAFFE, Two and three body equations in quantum field models, preprint.

[2] ________________, Three Particle structure of ϕ^4 Interactions and the Scaling Limit, Phys. Rev. D11, 2816 (1975)

[3] J. GLIMM, A. JAFFE, T. SPENCER, The particle structure of the weakly coupled $P(\Phi)_2$ model and other applications of high temperature expansions, parts I and II, in: Lecture Notes in Physics, Vol. 25, p. 132-242, Springer Verlag, Berlin-Heidelberg-New York, 1973.

[4] ____________________________, Ann. Math. 100, 585 (1974).

[5] K. HEPP, Commun. Math. Phys. 1, 95 (1965), see also in Axiomatic Field Theory, M. CHRETIEN and S. DESER (eds.), 1965 Brandeis Lectures, Vol. I, Gordon and Breach, New York - London-Paris, 1966.

[6] H. LEHMANN, K. SYMANZIK, W. ZIMMERMANN, Nuovo Cimento 1, 205 (1955); 6, 1122 (1951).

[7] K. OSTERWALDER, R. SCHRADER, Commun. Math. Phys. 31, 83 (1973); 42, 281 (1975).

[8] K. OSTERWALDER, R. SENEOR, The Scattering Matrix is Nontrivial for weakly coupled $P(\phi)_2$ Models. Preprint, to appear in Helv. Phys. Acta.

Acta Physica Austriaca, Suppl. XVI, 87–102 (1976)

Recent Progress on the $Yukawa_2$ Quantum Field Theory

Oliver A. McBryan*
Department of Mathematics
Rockefeller University
New York, N.Y. 1oo21

Abstract: In these talks, we review recent progress on the $Yukawa_2$ model for boson-fermion interactions in two space-time dimensions. We discuss in detail the Matthews - Salam formulation for integrating out the fermions, and obtain bounds on the resulting boson theory sufficient to show existence of an infinite volume limit. The vacuum energy density ("pressure") converges in this limit and we also obtain uniform ϕ-bounds which establish existence of infinite-volume Wightman functions.

From the point of view of mathematics, we study random compact operators $K(\phi)$-compact operators which are functions of a random process ϕ. We obtain bounds on the L_p norm of certain functionals of K, such as $\det(1+K)$, with respect to the underlying probability measure determined by ϕ.

The proof we present in section IV of bound on the partition function in a finite volume is considerably simpler that previous proofs and is particularly suited to small coupling constants.

*Research supported by the National Science Foundation under Grant MPS 74 - 13252

I. Introduction and General Survey

The Yukawa quantum field theory is a model for boson-fermion interactions with a trilinear interaction density, $\mathcal{H}_I(x) = \overline{\psi}\psi\phi(x)$. Progress on this model in two space-time dimensions (Y_2 model) has been quite rapid in the last year. Seiler [1], using formal ideas due to Matthews and Salam [2] showed that one can integrate out the fermions in a rigorous way. The model is then reduced to a purely boson problem, which allows for a euclidean treatment using techniques already developed for boson ($P(\phi)_2$) field theories. Previously, all results for the Y_2 model were obtained, with considerable effort, using non-covariant Hamiltonian methods. Because the Y_2 model is not so well-known as the $P(\phi)_2$ theories, I will begin by describing the principal results which are known for the model. Then in section II we describe some of the mathematical structures encountered in the euclidean approach. In section III the Matthews-Salam formalism is presented while in section IV we use it to obtain bounds on Schwinger functions suitable for controlling the infinite volume limit. In section V we discuss that limit, proving convergence of the energy density (pressure), and ϕ-bounds, and establishing existence of infinite volume Wightman functions as tempered distributions. An important ingredient is the use of Osterwalder-Schrader positivity to prove that the space of vacuum vectors for the finite volume Hamiltonian has nonzero overlap with the free vacuum.

The Y_2 model was first studied from a rigorous point of view by Glimm [3,4] and by Glimm and Jaffe [5,6] who showed existence of a semi-bounded self-adjoint finite volume Hamiltonian. Unlike $P(\phi)_2$ theories, existence of a finite volume Hamiltonian is non-trivial for the Y_2 model due to the presence of logarithmic ultraviolet divergences in the vacuum energy and boson mass. Schrader [7] used the finite volume Hamiltonians to show existence of an infinite volume limit in the sense of C^*-algebras, and

Glimm and Jaffe [8] proved locality of the local algebras. McBryan and Park [9] completed the proof of the Haag-Kastler axioms by demonstrating that the local algebras are Lorentz invariant.

There was little further progress beyond the Haag-Kastler axioms until the advent of euclidean techniques for Y_2. Osterwalder and Schrader [10] showed how to construct euclidean fermi fields. It is difficult to use these fields directly to obtain estimates because the corresponding euclidean action V_I is not a self-adjoint, or even normal, operator. Another possibility is however available - to integrate out the euclidean fermions. The possibility of doing this arises from the fact that the un-renormalized Hamiltonian $H = H_o + \lambda\int dx\overline{\psi}\psi\phi(x)$ is quadratic in the fermi fields. The corresponding field equations involve a pair of coupled linear equations for the fermions and thus, in principle, the fermions can be solved for and eliminated, yielding a more complex boson theory. Such a mechanism was first proposed over twenty years ago by Matthews and Salam [2] who used it for the case where the boson field is an external (classical) field. Seiler [1] has shown that this procedure also makes sense when the boson field is itself quantized. Specifically, he has shown that the Schwinger functions (euclidean Green's functions) for the Yukawa$_2$ model with a space-time cutoff, but no momentum cut-offs, may be expressed purely in terms of expectation values of a boson field with respect to a complicated non-local interaction. Following this development there has been rapid progress. Bounds on finite volume Schwinger functions suitable for controlling the infinite volume limit were proved by McBryan [11], Seiler and Simon [12] and Magnen and Seneor [13]. More recently, convergence of the vacuum energy density and of the euclidean pressure as the volume tends to infinity, as well as uniform ϕ-bounds, have been proved by McBryan [14] and by Simon and Seiler [15]. An immediate consequence of the ϕ bounds is the existence of infinite volume Schwinger functions as tempered distributions [16]. Finally, while all of the above results apply to arbitrary coupling strength, in the case of sufficiently weak coupling Magnen and Seneor [13] have announced a cluster expansion which establishes all of the Wightman axioms for the infinite volume theory. Since that will be the content of R. Seneor's talks, I will not discuss the cluster expansion further in these lectures.

II. Mathematical Structures

From the point of view of mathematics, the problems we will discuss for the Y_2 model involve random compact operators - operators $K(\phi)$ where K is an almost everywhere defined non-normal compact operator depending on a random process ϕ. We will be interested in quantities such as

$$\int d\mu_o \overset{m}{\underset{a}{\otimes}} (1+K(\phi))^{-1} \det(1+K(\phi)). \tag{2.1}$$

Here $(d\mu_o, \phi(\ ))$ is the Gaussian process indexed by $\mathcal{H}_{-1}^{(\mu)}(R^2)$, where $\mathcal{H}_s^{(\mu)}(R^2) \equiv L_2(R^2; (k^2+\mu^2)_h\ d^2k)$, while $\overset{m}{\underset{a}{\otimes}}$ denotes an m-fold antisymmetric tensor product of operators. Thus $(d\mu_o, \phi(\cdot))$ is the Gaussian process with mean zero and covariance $-\Delta+\mu^2$, [17,18].

A serious problem arises because the operators $K(\phi)$ turn out not to belong to either the trace class or Hilbert-Schmidt ideals. This poses difficulties in defining the determinants in (2.1) since the determinant of an operator exists only for trace class operators. The resolution of this problem, which in physics is described as renormalization, is to define regularized determinants

$$\det_p(1+K) \equiv e^{\mathrm{Tr}\left[-K+\frac{1}{2}K^2 \ldots + (-)^{p-1}K^{p-1}\right]}\det(1+K) \tag{2.2}$$

For $K \in C_p$ - the ideal of compact operators such that $||K||_p = (\mathrm{Tr}(K^+K)^{p/2})^{1/p}$ is finite - the quantity $\det_p(1+K)$ is a well-defined analytic function of K, see for example Dunford and Schwartz [19].

To determine if $K(\phi)$ is in one of the classes C_p we need estimates on the rate of decrease of the ordered eigenvalues $\mu_n(\phi)$ of $|K(\phi)|$ as $n \to \infty$. Up to multiplication by a unitary operator, the operator $K(\phi)$ has the form of an integral operator on $L_2(R^2)$ with kernel

$$K_\Lambda(\phi) = (-\Delta + m^2)^{-1/2} \chi_\Lambda \phi \,. \tag{2.3}$$

Here χ_Λ is the characteristic function of a compact region $\Lambda \subset R^2$, Δ is the Laplacian and we regard $\chi_\Lambda\phi$ as a multiplication operator. One can show by estimates on the eigenvalues [20,21] that almost everywhere with respect to $d\mu_o$, $K_\Lambda(\phi)$ belongs to C_p, $p > 2$, and thus $\det_3(1+K_\Lambda(\phi))$ is well-defined. The principal technical estimate we require in order to establish the results mentioned in section I, will be an estimate of the form:

$$\int d\mu_o || \bigotimes_a^n (1+K_\Lambda(\phi))^{-1} || \det_3(1+K_\Lambda(\phi)) \leq c_1^n \, c_2^{|\Lambda|}, \tag{2.4}$$

where $|\Lambda|$ is the area of Λ and c_1, c_2 are constants.

To prove integrability of $\det_3(1+K_\Lambda(\phi))$ we will use a Taylor's series for $\det_3$ to order $n(\phi)$ depending on the magnitude of the random variable ϕ and we estimate the remainder of ϕ, by the Hadamard inequality for determinants which uses the cancellations inherent in the determinant. The integral is then bounded by estimating the measure of the sets on which ϕ attains large values. To obtain the more detailed bound (2.4) we write $K_\Lambda(\phi)$ as a sum of operators localized in unit squares, perform a separate Taylor expansion for each square, and use the fact that random variables $\phi(x)$, $\phi(y)$ are almost independent for large separation $|x-y|$.

In section III we indicate how the Yukawa_2 model leads to the study of quantities like (2.1), (2.3), and in section IV we prove a bound of the form (2.4) by a method similar to that indicated in the previous paragraph.

III. The Matthews-Salam Formalism

III.1 Euclidean Fermi Fields

Before discussing the Matthews-Salam formalism, I would like to say something about the euclidean fermi fields which have been constructed by Osterwalder and Schrader [10]. Suitable candidates for euclidean fields should be anti-commuting free fields $\Psi(x)$ such that

$$\left\langle \Psi_\alpha(x)\Psi_\beta(y)^* \right\rangle = C\left\langle T\Psi_\alpha(x)\Psi_\beta^*(y)\right\rangle \equiv S_{o,\alpha,\beta}(x,y),$$

where C denotes the analytic continuation from $x_o \to ix_o$, $y_o \to iy_o$, T denotes time ordering and ψ_α is the free relativistic fermi field of mass m. However it is not possible to satisfy this relation since the right-hand side is not hermitean. This difficulty seems to be avoidable only by the device of introducing separate euclidean fermi fields $\Psi^{(i)}$, i=1,2, corresponding to Ψ and $\bar{\Psi}$. Thus we choose fields such that

$$\left\langle \Psi^{(1)}(x)\Psi^{(2)}(y)\right\rangle = S_o(x,y) = (2\pi)^{-2}\int d^2p e^{ip\cdot(x-y)}(\not{p}-m)^{-1}$$

$$\left\langle \Psi^{(i)}(x)\Psi^{(i)}(y)\right\rangle = 0 \quad , \quad i = 1,2. \tag{3.1}$$

The euclidean action, partition function and un-normalized Schwinger functions are:

$$V_I(\Lambda) \equiv \int_\Lambda d^2x\phi(x)\Psi^{(2)}(x)\Psi^{(1)}(x) \quad , \quad Z^{(\Lambda)} \equiv \left\langle e^{-\lambda V_I(\Lambda)}\right\rangle \tag{3.2}$$

$$(ZS)^{(\Lambda)}_{n,m}(x;y;z) \equiv \left\langle \prod_{i=1}^{n}\phi(x_i)\prod_{j=1}^{m}\Psi^{(1)}(y_j)\prod_{k=1}^{m}\Psi^{(2)}(z_k)e^{-\lambda V_I(\Lambda)}\right\rangle$$

Here < > denotes expectation in the tensor product Ω of the vacuum Ω_b for the boson field and the vacuum Ω_f for the fermion fields. As we will see shortly, none of these quantities is well-defined without renormalizations. For further details concerning euclidean fermi fields and the proof of a Feyman-Kac formula relating euclidean Schwinger functions to time ordered products of relativistic fields, we refer to the paper of Osterwalder and Schrader [10].

III.2 Integrating out the Fermions

In order to integrate out the fermions, we will consider the expectations in (3.2) only for the fermion vacuum state (which we denote by $\langle\ \rangle_f$), regarding the boson field as a classical quantity which can later be integrated over. For simplicity, we consider first the zero and two-point functions

$$Z_f = \left\langle e^{-\lambda V_I}\right\rangle_f \quad , \quad S_f(y,z) = \left\langle \psi^{(1)}(y)\psi^{(2)}(z)e^{-\lambda V_I}\right\rangle_f \quad ,$$

where we have suppressed the dependence on the interaction volume Λ in our notation. Decomposing $\psi^{(1)}(y)$ into a creation and an annihilation part, and contracting the latter to the right, we obtain

$$\begin{aligned} S_f(y,z) &= S_o(y,z)\left\langle e^{-\lambda V_I}\right\rangle_f - \left\langle \psi^{(2)}(z)\left[\psi^{(1)-}(y),\lambda V_I(\Lambda)\right]e^{-\lambda V_I}\right\rangle_f \\ &= S_o(y,z)Z_f - \lambda\int_\Lambda d^2x S_o(y,x)\phi(x)\left\langle\psi^{(1)}(x)\psi^{(2)}(z)e^{-\lambda V_I}\right\rangle_f \quad . \end{aligned} \tag{3.3}$$

Introducing the integral operator $K = K_\Lambda(\phi)$ with kernel

$$K(x,y) = S_o(x,y)(\chi_\Lambda\phi)(y) \quad ,$$

where χ_Λ is the characteristic function of Λ, we obtain from (3.3)

$$S_f(y,z) = ((1+\lambda K)^{-1}S_o)(y,z)Z_f \; . \tag{3.4}$$

To compute Z_f we differentiate $Z_f(\lambda)$ with respect to λ:

$$\frac{\partial}{\partial\lambda} Z_f(\lambda) = \left\langle -V_I e^{-\lambda V_I}\right\rangle_f = \int_\Lambda d^2x\phi(x)S_f(x,x)$$

$$= \int d^2x(\chi_\Lambda\phi)(x)((1+\lambda K)^{-1}S_o)(x,x)Z_f(\lambda)$$

$$= Z_f(\lambda)\mathrm{Tr}K(1+\lambda K)^{-1}$$

where in the third step we used (3.4). Thus, on integrating,

$$Z_f(\lambda) = e^{\mathrm{Tr}\ln(1+\lambda K)} \equiv \det(1+\lambda K)^{+} \qquad (3.5)$$

From (3.4), (3.5) we obtain $S_f(y,z)$ explicitly in terms of K.

Renormalization of λV_I requires that we subtract (i) $\lambda V_I = \mathrm{Tr}\lambda K$ which normal orders V_I, (ii) a vacuum energy counter-term $-\frac{1}{2}\langle:\lambda V_I:^2\rangle = \frac{1}{2}\langle \mathrm{Tr}\lambda^2K^2\rangle_b$ and (iii) a boson mass counter-term $-\frac{1}{2}\lambda^2\delta\mu^2:\phi^2(\chi_\Lambda):$, $\delta\mu^2 = -2(2\pi)^{-2}\int d^2p(p^2+m^2)^{-1}$. Thus we must replace $Z_f = \det(1+\lambda K)$ by

$$\det{}_{ren}(1+\lambda K) \equiv e^{-\lambda \mathrm{Tr}K+\frac{1}{2}\lambda^2<\mathrm{Tr}K^2>_b+\frac{1}{2}\lambda^2\delta\mu^2:\phi^2(\chi_\Lambda):}\det(1+\lambda K)$$

$$= e^{-\frac{1}{2}\lambda^2:F(\phi):}\det{}_3(1+\lambda K) \qquad (3.6)$$

where $F(\phi) \equiv -\mathrm{Tr}K^2 + \delta\mu^2\phi^2(\chi_\Lambda)$. As indicated in section II, the occurrence of $\det_3(1+\lambda K)$ is expected on mathematical grounds because K is (a.e. with respect to $d\mu_o$) in C_p for $p > 2$ - specifically one shows that $\int d\mu_o \mathrm{Tr}(K^*_\Lambda(\phi)K_\Lambda(\phi))^{p/2}$ is finite for $p > 2$.

Instead of considering only the two-point function as above, we can apply a similar analysis to the 2m-point fermi expectation $S_f^{(m)}(y;z) \equiv \langle \Psi^{(1)}(y_1)\ldots\Psi^{(1)}(y_m)\Psi^{(2)}(z_1)\ldots\Psi^{(2)}(z_m)e^{-\lambda V_I}\rangle_f$ reducing it to $S_f^{(m-1)}$ and thus by induction to Z_f. Thus for the renormalized fermi expectation values we obtain:

+For a self-adjoint trace-class operator A with eigenvalues λ_i, $\det(1+A) \equiv \prod_i(1+\lambda_i) = \exp(\sum_i \ln(1+\lambda_i)) = \exp(\mathrm{Tr}\ln(1+A))$
Generalization to non-selfadjoint A is not difficult [19].

$$S_f^{(m)}(y;z) = (-)^{[m/2]} \det_{jk}((1+\lambda K)^{-1} S_o)(y_j, z_k) \det_{ren}(1+\lambda K) \tag{3.7}$$

where $\det_{jk} a_{jk}$ denotes the m x m determinant with elements a_{jk}.

IV. Bounds on Schwinger Functions

In this section we will indicate the proof of bounds on un-normalized Schwinger functions

$$(ZS)^{(\Lambda)}_{n,m}(f;g;h) \equiv \left\langle \prod_{i=1}^{n} \phi(f_i) \prod_{j=1}^{m} \Psi^{(1)}(g_j) \prod_{k=1}^{m} \Psi^{(2)}(h_k)\, e^{-V(\Lambda)} \right\rangle$$

and on the partition function $Z^{(\Lambda)} \equiv (ZS)^{(\Lambda)}_{(o,o)}$. Here f_i, g_j, h_k are functions in the boson or fermion test function spaces $\mathcal{H}^{(\mu)}_{-1}$ or $\mathcal{H}^{*} = \mathcal{H}^{(m)}_{-1/2} \oplus \mathcal{H}^{(m)}_{-1/2}$, with supports in unit squares belonging to the lattice of unit squares Δ_α with centers $\alpha \in Z^2$, while $V(\Lambda)$ denotes the Y_2 interaction renormalized as in section III.2. Thus by (3.7):

$$(ZS)^{(\Lambda)}_{n,m}(f;g;h) = (-)^{[m/2]} \int d\mu_o \prod_{i=1}^{n} \phi(f_i) \det_{jk} S_F(g_j, h_k) \det_{ren}(1+\lambda K) \tag{4.1}$$

with $K = K_\Lambda(\phi)$ defined previously and $S_F(g,h) \equiv (g,(1+\lambda K)^{-1} S_o h)_{L_2}$. Let n_α denote the number of boson test functions localized in square Δ_α. Then

<u>Theorem 4.1</u> There are constants $c_1, c_2, c_3 \neq o$, independent of n,m,Λ, with

(i) $|(ZS)^{(\Lambda)}_{n,m}(f;g;h)| \leq c_1^{|\Lambda|} c_2^{n+m} \prod_\alpha n_\alpha!^{1/2} \prod_{i=1}^{n} ||f_i||_{-1} \prod_{j=1}^{m} ||g_j||_{-1/2}$

$||h_j||_{-1/2}$

(ii) $Z^{(\Lambda)} \geq c_3^{|\Lambda|}$, if Λ is a rectangle.

For a complete proof of Theorem 4.1 we refer to McBryan [11]; the estimate (i) has been proved also by Seiler and Simon [12]. We will indicate the main ideas by giving a simplified proof of (i), for the case $m = o$, $|\Lambda| = 1 = \lambda$. Applying the Schwartz inequality to (4.1), to separate $\prod_{i=1}^{n} \phi(f_i)$, we obtain the terms

$c_2^n \prod_\alpha n_\alpha!^{1/2} \prod_i ||f_i||_{-1}$ by a bound on the free field L_2 norm of $\prod_{i=1}^n \phi(f_i)$. Thus it remains only to show that $\det_{ren}(1+K) \in L_2(d\mu_o)$. This was essentially the content of Seiler's original paper [1], but the proof we give now is considerably shorter.

We introduce cutoff fields and operators, ϕ_n and $K_n = K(\phi_n)$ where $\phi_n = \chi_{\kappa_n} * \phi$, with $\chi_\kappa(x) \equiv \kappa^2 \chi(\kappa x)$, $\chi \in C_o^\infty(R^2)$, $\int d^2x \chi(x) = 1$, and with $\kappa_n = {}^n\mu(e^n - 1)$. We use the identity:

$$\det{}^2_{ren}(1+K(\phi)) = 1 + \sum_{n=1}^\infty \{\det{}^2_{ren}(1+K_n) - \det{}^2_{ren}(1+K_{n-1})\} \qquad (4.2)$$

To estimate the differences, we require interpolating fields $\phi_{n(s)} \equiv s\phi_n + (1-s)\phi_{n-1}$ and operators $K_{n(s)} \equiv K(\phi_{n(s)})$, so that

$$\det{}^2_{ren}(1+K_n) - \det{}^2_{ren}(1+K_{n-1}) = \int_o^1 ds \frac{\partial}{\partial s} \det{}^2_{ren}(1+K_{n(s)})$$

$$= \int_o^1 ds e^{-:F(\phi_{n(s)}):} \det{}^2_3(1+K_{n(s)}) \{-:\frac{\partial}{\partial s} F(\phi_{n(s)}): + \frac{2}{3} \mathrm{Tr}(1+K_{n(s)})^{-1} \frac{\partial K^3_{n(s)}}{\partial s}\}$$

The integrand may be bounded using $|\mathrm{Tr}AB| \leq ||A|| \, ||B||_1$ and the

Lemma: For A Hilbert Schmidt, $|\det_3(1+A)|$ and $||(1+A)^{-1}\det_3(1+A)||$ are bounded by $e^{1+\frac{1}{2}\mathrm{Tr}(A^2+A^*A)}$.

Both of these bounds follow from applying $|(1+\lambda)e^{-\lambda}| \leq e^{|\lambda^2|/2}$ to each eigenvalue factor in $\det_3(1+A)$. Thus we obtain

$$|\det{}^2_{ren}(1+K_n) - \det{}^2_{ren}(1+K_{n-1})| \qquad (4.3)$$
$$\leq e \sup_s e^{\langle H(\phi_{n(s)})\rangle - :G(\phi_{n(s)}):} \{|:\frac{\partial F}{\partial s}(\phi_{n(s)}):| + ||\frac{\partial K^3_n}{\partial s}(s)||_1\}$$

where $G(\phi) \equiv \mathrm{Tr}K^*(\phi)K(\phi) + \delta\mu^2\phi^2(\chi_\Lambda)$, $H(\phi) \equiv \mathrm{Tr}(K^*(\phi)K(\phi)+K^2(\phi))$, and to obtain an estimate on the L_1 norm of (4.3) we integrate with respect to $d\mu_o$ and apply the Schwartz inequality on the right. Since both $G(\phi)$ and $H(\phi)$ are quadratic functions of ϕ, $||\exp(-:G(\phi_{n(s)}):)||_2$ and $\langle H(\phi_{n(s)})\rangle$ may be explicitly evaluated, and bounded respectively by const and by const $\ln(\kappa_n/\mu)$, with constants uniform in n, s, [1,11]. Furthermore both $|| :\frac{\partial F}{\partial s}(\phi_{n(s)}): ||_2$ and $|||| \frac{\partial K^3_{n(s)}}{\partial s} ||_1||_2$ are easily bounded by const $\kappa_{n-1}^{-\varepsilon}$, with $\varepsilon>o$ and const uniform in n, s, [11]. These bounds are related to the fact that $:F(\phi_n):$ and K_n^3 converge a.e. to $:F(\phi):$ and K^3, respectively (the K^3 convergence is in the trace norm), and follow from the corresponding Feynman integrals.

Inserting the resulting bound for (4.3) into (4.2) gives

$$\begin{aligned}\int d\mu_o \det{}^2_{ren}(1+K(\phi)) &\leq \text{const}(1+\sum_{n=2}^{\infty} e^{\text{const}\, \ln(\kappa_n/\mu)}\, \kappa_{n-1}^{-\varepsilon} \\ &\leq \text{const}(1+\sum_{n=2}^{\infty} e^{\text{const}\, n}\, e^{-\varepsilon n}) \\ &< \infty\,, \qquad (4.4)\end{aligned}$$

provided const in the exponent is sufficiently small $(<\epsilon)$.$^+$ The constant in the exponent may be made small by removing from V_I the contribution from fermions with low momenta, say less than $\rho = \rho(\varepsilon)$. The bound (4.4) applies with K replaced by K_ρ, the operator corresponding to the remaining high momentum interaction. However, as we have shown in [20], the low-momentum part of the interaction in such a situation may be easily replaced (because cutoff fermions are bounded operators), the only effect being to increase the bound (4.4) further.

To obtain the volume dependance $c_1^{|\Lambda|}$ in Theorem 4.1, one carries out the above expansion repeatedly in each unit volume contained in Λ, noting that the correlation between distant volumes is exponentially small. For the details we refer to McBryan [11]. This completes our discussion of Theorem 4.1.

$^+$ The constant in the exponent is proportional to λ^2 when K is replaced by λK. Thus for small coupling the bound (4.4) follows immediately. The argument of the remainder of the paragraph is required only for large $|\lambda|$.

V. Applications

The bounds of Theorem 4.1 are sufficient to control the infinite volume limit of the Yukawa_2 model. Let H_L denote the self-adjoint Hamiltonian for space-volume $[-L/2\ ,\ L/2]$ which has been constructed by Glimm and Jaffe, [5]. In particular they have shown that the lowest point in the spectrum of H_L is an eigenvalue E_L of finite multiplicity. In order to study E_L, we will use the Feyman-Kac formula proved by Osterwalder and Schrader [10], which connects the euclidean and relativistic Fock spaces. After renormalization as in section III.2, it takes the form, [11,14]:

$$<\Omega_o, e^{-tH_L}\Omega_o> = \int d\mu_o \det{}_{ren}(1+K_{\Lambda_{t,L}}(\phi))e^{-W(t,L)} \tag{5.1}$$

where $\Lambda_{t,L} = [-t/2,t/2]\times[-L/2,L/2]$, $W(t,L) = \langle H_{I,L}H_o^{-2}(1-e^{-tH_o})$ $H_{I,L}\rangle$, and we have taken $\lambda = 1$ for convenience. Let P_L denote the projection onto vacuum vectors for H_L - i.e. onto eigenvectors for the eigenvalue E_L. A crucial step required before we can use (5.1) to control E_L is to know that the vacuum subspace for H_L is not orthogonal to Ω_o:

Theorem 5.1 $P_L\Omega_o$ is nonzero for each finite L.

For the proof of Theorem 5.1 we refer to our contribution to the Marseille conference [14]. We use only Osterwalder-Schrader positivity in the space direction and Nelson's symmetry applied twice. Seiler and Simon have also announced all of the results of this section [15], but we have not yet seen their proofs. An immediate consequence of Theorem 4.1 and 5.1 is Schrader's linear bound on the vacuum energy [7]:

Theorem 5.2 $|E_L| \leq \text{const } L$, if $L \geq 1$.

Proof By (5.1) and Theorem 5.1:

$$E_L = -\lim_{t\to\infty} t^{-1} \ell n \langle \Omega_o, e^{-tH_L}\Omega_o \rangle \tag{5.2}$$

$$= -\lim_{t\to\infty} t^{-1} \ell n \int d\mu_o \det{}_{ren}(1+K_{\Lambda_{t,L}}) + \lim_{t\to\infty} t^{-1} W(t,L)$$

The first term is bounded in magnitude by const L by Theorem 4.1, while the second term converges to zero as $t \to \infty$.

We will next study the vacuum energy density $\alpha_L = -E_L/L$ by the previous theorem is bounded uniformly in L.

<u>Theorem 5.3</u> α_L converges to a finite constant α_∞ as $L \to \infty$.

<u>Proof</u> We will show that for a constant $c > o$, and any $o<a<1$:

$$-E_{aL} \le -aE_L + (1-a)c$$

It follows that $\alpha'_L \equiv -(E_L+c)/L$ satisfies

$$\alpha_{aL} \le \alpha'_L \quad , \quad o<a<1, \text{ all } L, \quad \alpha'_L = \alpha_L - c/L \le \sup_L \alpha_L < \infty$$

Thus α'_L is monotonically convergent to a finite limit α_∞ and clearly $\alpha_L = \alpha'_L + c/L$ also converges to this limit.

To prove (5.3) we use (5.2) and the Feynman-Kac formula (5.1). For convenience denote $\int d\mu_o \det_{ren}(1+K_{\Lambda_{t,L}})$ by $Z_{t,L}$. Then

$$\begin{aligned}
(\Omega_o, e^{-tH_{aL}}\Omega_o) &= Z_{t,aL} e^{-W(t,aL)} \\
&= Z_{aL,t} e^{-W(t,aL)} \\
&= (\Omega_o, e^{-aLH_t}\Omega_o) e^{-W(t,aL)+W(aL,t)} \\
&\le (\Omega_o, e^{-LH_t}\Omega_o)^a e^{-W(t,aL)+W(aL,t)} \\
&= Z^a_{L,t} e^{-W(t,aL)+W(aL,t)-aW(L,t)} \\
&= Z^a_{t,L} e^{-W(t,aL)+W(aL,t)-aW(L,t)} \\
&= (\Omega_o, e^{-tH_L}\Omega_o)^a e^{-W(t,aL)+aW(t,L)+W(aL,t)-aW(L,t)}
\end{aligned}$$

where we have used the inequality $(\theta,A^a\theta) \leq (\theta,A\theta)^a$, $o<a<1$, as well as Nelson's symmetry for $Z_{t,L}$. Thus

$$-E_{aL} \leq -aE_L+\lim_{t\to\infty} t^{-1}\{-W(t,aL)+aW(t,L)+W(aL,t)-aW(L,t)\}.$$

Since $W(t,L)$ is bounded uniformly in t, the first two terms in brackets do not contribute. For the other two terms note that:

$$W(aL,t)-aW(L,t) = \left\langle H_{I,t}H_o^{-2}(1-a+ae^{-LH_o}-e^{-aLH_o}),H_{I,t}\right\rangle$$

$$\leq (1-a)\left\langle H_{I,t}H_o^{-2}H_{I,t}\right\rangle \leq (1-a)ct.$$

The inequality (5.3) now follows immediately.

As a final application of these techniques, we discuss the ϕ-bound for Yukawa_2:

<u>Theorem 5.4</u> Let $f\in C_o^\infty(R^1)$, support $f\in(-L/2,L/2)$, $L\geq 1$. Then for a contant C and a Schwartz space norm $|||\cdot|||$, independent of L,f:

$$\pm\phi_o(f) \leq C|||f|||(H_L - E_L + 1)$$

An immediate consequence is the existence of infinite volume Wightman functions as tempered distributions (note that the fermion operators are bounded), [16]. The proof follows essentially the ideas of Guerra, Rosen and Simon for $P(\phi)_2$ models [18,22]. For details we refer again to our Marseille talk [14]. The main inputs are the convergence and monotonicity of α_L', Nelson's symmetry and Osterwalder-Schrader positivity.

References

[1] SEILER, E., Commun. Math. Phys. 42, 163 - 182 (1975)

[2] SALAM, A., and MATHEWS, P. T., P.R. 90, 690 - 695 (1953)

[3] GLIMM, J., Commun. Math. Phys. 5, 343 - 386 (1967)

[4] ——— Commun. Math. Phys. 6, 120 - 127 (1967)

[5] GLIMM, J. and JAFFE, A., Ann. of Phys. 60, 321 - 383 (1970)

[6] ——— Field Theory Models, in Statistical Mechanics and Quantum Field Theory, edited by C. Dewitt and R. Stora, Gordon and Breach, New York 1971.

[7] SCHRADER, R., Ann. of Phys. 70, 412 - 457 (1972)

[8] GLIMM, J. and JAFFE, A., J. Funct. Analysis 7, 323 - 357 (1971)

[9] MCBRYAN, O. and PARK, Y. M., J. Math. Phys. 16, 104 - 110 (1975)

[10] OSTERWALDER, K. and SCHRADER, R., Helv. Phys. Acta 46, 277 - 302 (1973)

[11] MCBRYAN, O., Volume dependance of Schwinger functions in the $Yukawa_2$ quantum field theory, Commun. Math. Phys. 45, 279 - 294 (1975)

[12] SEILER, E. and SIMON, B., Bounds in the $Yukawa_2$ quantum field theory, Commun. Math. Phys., to appear.

[13] MAGNEN, J. and SENEOR, R., Announced at this meeting.

[14] MCBRYAN, O., Convergence of the Vacuum Energy Density, ϕ-bounds and Existence of Wightman Functions for the $Yukawa_2$ Model, to appear in Proceedings of the International Colloquium on Quantum Fields, C.N.R.S., Marseille, June 1975.

[15] SEILER, E., and SIMON, B., Announcement, Private Communications.

[16] GLIMM, J. and JAFFE, A., J. Math. Phys. 13, 1568 - 1584 (1972)

[17] NELSON, E., J. Funct. Anal. 12, 211 - 227 (1973)

[18] SIMON, B., The $P(\phi)_2$ Euclidean Field Theory, Princeton University Press, Princeton 1974.

[19] DUNFORD, N. and SCHWARTZ, J., Linear Operators, Part II: Interscience Publishers, New York 1963.

[20] MCBRYAN, O., Commun. Math. Phys. 44, 237 - 243 (1975)

[21] SEILER, E. and SIMON, B., On finite mass renormalizations in the two dimensional Yukawa model, J. Math. Phys. to appear.

[22] GUERRA, F., ROSEN, L., and SIMON, B., Commun. Math. Phys. 27, 10 - 22 (1972)

Acta Physica Austriaca, Suppl. XVI, 103–124 (1976)

The Cluster Expansion for Y_2

R. SENEOR
Centre de Physique Théorique
E. Polytechnique,
Palaiseau,
France

I. Introduction

For long time, the results concerning the Euclidean Yukawa quantum field theory (Y_2) in two dimension were very far from the ones obtained in $P(\Phi)_2$ quantum field theories. This difference was essentially due to the difficulties one has in describing Euclidean Fermi fields. However since the definition of this model solely in term of bose fields (i.e. with the fermions "integrated out") given by E. SEILER [1] considerable progress has been made. Upper bounds depending exponentially on the interaction volume as in $P(\Phi)_2$ have been obtained by O.McBRYAN [2] and E. SEILER and B. SIMON [3]. The comparison with $P(\Phi)_2$ theories is even more complete since McBRYAN [4] has obtained the proof of Φ-bounds, and by the way, of the existence of Wightman functions. The next step to complete the analogy with $P(\Phi)_2$ theories was to prove the convergence of a cluster expansion. This is the result obtained in collaboration with J. MAGNEN [5] that I will report here.

The definitions, which are slightly different than the usual ones, will be given in part II. The cluster expansion is then presented in part III, where also the main results are stated. The sketches of the proofs are then given in parts IV and V.

II. Definitions

The partition function of the model, for an interaction in the volume Λ, is given by

$$Z_\Lambda = \int d\mu \ \det_{ren} (1+K_\Lambda) \qquad \text{(II.1)}$$

where

$$K_\Lambda(x,y) = S_m(x,y)\Gamma\Phi(y)\ \Lambda(y)$$

S_m (x,y) being the free fermion covariance, $\Lambda(y)$ being the characteristic function of the volume Λ, λ being the coupling constant and $\Gamma=1$, the scalar case or $\Gamma=i\gamma_5$, the pseudoscalar case. In the sequel, we consider only the scalar case. By definition

$$\det_{ren} (1+K_\Lambda) = \det_3 (1+K_\Lambda)\ B_M$$

where $\det_3$ (1+K) is defined as in McBryan's lecture and

$$B_M = e^{-\frac{1}{2} \mathrm{Tr}_{reg} :K_\Lambda^2: -\frac{1}{2} M^2 \int :\phi^2:(x)\Lambda(x)dx} \qquad \text{(II.2)}$$

Remark that this counterterm is different from the one defined by McBryan, but has the form given in the original Seiler's article [1]. We will comment on this fact in the last part.

Now for test functions $f_1,\ldots,f_n,g_1,\ldots,g_N,h_1,\ldots,h_N$ in a suitable space, the unnormalized Schwinger functions are given by

$$S_\Lambda\ (f_1,\ldots f_n;\ g_1,\ldots,g_N;\ h_1,\ldots,h_N) =$$

$$= \int d\mu \, \det_{i,k} S_F \, (g_i, h_k; \Phi) \prod_{i=1}^{N} \Phi \, (f_i) \, \det_{ren} \, (1+K_\Lambda) \qquad \text{(II.3)}$$

$\det_{i,k} S_F$ is defined as in McBryan's lecture, and

$$S_F = \frac{1}{1+K_\Lambda} S_m$$

Remark that we have not precised the covariance associated with the boson field Φ. Indeed we take

$$C_m(x,y) = (-\Delta+m^2)^{-1}(x,y)$$

with the same mass as for the fermions. This equality of the values of the masses is only for simplicity, since as we will see later, the parameter which will force the convergence of the cluster expansion is the infimum of these two masses.

In order to get bounds on these expressions we need to define a topology for the Fredholm operator K_Λ. At this point we differ from Seiler's and McBryan's definitions and I want to explain why.

We consider K as an operator in $\mathcal{H}_{1/2} = \mathbb{C}^2 \otimes L_2(F_m^{-1/2}(p)d^2p)$ instead of $\mathbb{C}^2 \otimes L_2 \, (\mu_m(p)d^2p)$ where $F_m(p)$ is the comparison function introduced by J. Glimm and A. Jaffe [6].

$$F_m(p) = \prod_{i=0}^{1} \mu_m^{-1}(p_i) \text{ and } \mu_m(p) = \sqrt{p^2+m^2}$$

and p^2 is the Euclidean norm of $p = (p_0, p_1) \in \mathbb{R}^2$.

Before I justify this replacement of $\mu_m(p)$ by $F_m^{-1/2}(p)$, let me first explain why to take $\mathcal{H}_{1/2}$ instead of $\mathcal{H}_\alpha$ (with an obvious definition) for some other value of α.
In $\mathcal{H}_\alpha$, the kernel $K(p,q)$ of K is given by*

* By the natural isomorphism between $\mathcal{H}_\alpha$ and $\mathbb{C}^2 \otimes L_2(d^2p)$ it is always possible to consider K as an operator in this last space with kernel given by

$$\hat{K}(p,q) = F_m(p)^{-\alpha/2} \, \tilde{S}_m(p) \, \tilde{\Phi} \, (p-q) \, F_m(q)^{\alpha/2}$$

$$K(p,q) = \tilde{S}_m(p)\ \tilde{\Phi}\ (p-q)\ F_m(q)^\alpha$$

and $K^*(p,q) = \overline{K(q,p)}^{tr}$.
Now in estimating we will get expressions containing KK^* or K^*K. But

$$(KK^*)(p,q) = \int \tilde{S}_m(p)\ \tilde{\Phi}\ (p-t)\ F(t)^\alpha \tilde{\Phi}\ (t-q)\ \tilde{S}_m\ (-q)\ dt$$

$$(K^*K)(p,q) = F_m(p)^\alpha \tilde{\Phi}(p-t)\ \tilde{S}_m(-t)\ F_m(t)^{-\alpha} \tilde{S}_m(t) \tilde{\Phi}(t-q)\ F(q)^\alpha dt$$

We thus see that apart the usual propagator $\tilde{S}_m(p)$ these expressions generate between Φ-fields two new kinds of propagators $F_m(t)^\alpha$ and $\frac{1}{t^2+m^2} F_m(t)^{-\alpha}$.
Using now the following inequality

$$\mu_m^{-2}(p) \leq F_m(p) \leq \frac{1}{m} \mu_m^{-1}(p) \qquad \text{(II.4)}$$

and the fact that each matrix element of $\frac{\not{p}+m}{p^2+m^2}$ is bounded by $2\mu_m^{-1}(p)$, one sees that the fermion propagators are bounded by $F_m(p)^{1/2}$ for the usual one and by $F_m(p)^\alpha$ and $F_m(p)^{1-\alpha}$ for the two other ones.

Therefore a natural choice is to take $\alpha = 1/2$. We would have made the same choice if instead of $F_m(p)^\alpha$ we would have taken $\mu_m(p)^{-2\alpha}$ so, it remains to justify the choice of F instead of μ. The reason is the following. At the end of many procedures in Euclidean Quantum Field Theory, bounds reduce to the computation of fully contracted expressions, i.e. graphs, and we have chosen to compute these graphs by localizing each vertices. To sum up these localized graphs we need to extract localization factors (see [6]) and we are lead finally to bound expressions containing derivatives of the Fourier transforms of characteristic functions. But a natural bound for these derivatives is given by the comparison function:

$$\left|\left(\frac{\partial}{\partial p}\right)^\alpha \tilde{\chi}\ (p)\right| \leq O(1)\ F(p)$$

Thus we see that due to our way of estimating graphs, we have to replace quantities in term of μ by quantities in term of F using inequality (II.4). Unfortunately the second member of (II.4) is not optimal since one bounds an almost L_1 function by an almost L_2 one. This is roughly speaking the motivation of our choice.

Remark however that this modification is purely technical and can be probably removed by using finer local estimates as in |3|.

III. The Cluster Expansion: The Framework

The cluster expansion for Y_2 is developped along the same lines as for the $P(\Phi)_2$ theories.

One introduces a lattice cover of $\mathbb{R}^2$ with unit squares: $\mathbb{R}^2 = \bigcup_{\alpha \in Z^2} \Delta_\alpha$, where Δ_α is the unit square centered at α. One call Z^{2*} the set of all lattice lines or bonds.

Roughly speaking the principle of the cluster expansion is to compare the initial expression where all squares are coupled with expressions where the coupling between some (or all) squares has been removed. This is obtained by replacing the free covariances C and S by covariances C_Γ and S_Γ corresponding to Green's functions with Dirichlet boundary conditions on $\Gamma \subset Z^{2*}$. To compare the different expressions corresponding to these modified covariances, a variable s_b, $o \leq s_b \leq 1$ is associated to each bond $b \in Z^{2*}$, in order to interpolate between Dirichlet and no boundary conditions. One defines in this way

$$C_{(s)}(x,y) = \sum_{\Gamma \in Z^{2*}} \prod_{b \in \Gamma} s_b \prod_{b \notin \Gamma} (1-s_b)\, C_\Gamma(x,y)$$

$$S_{(s)}(x,y) = \sum_{\Gamma \subset Z^{2*}} \prod_{b \in \Gamma} s_b \prod_{b \notin \Gamma} (1-s_b)\, S_\Gamma(x,y)$$

and $Z_{(s),\Lambda}$ and $S_{(s),\Lambda}(f_1,..,f_n;\, g_1,..,g_N;\, h_1,...,h_N)$ where in (II.1) and (II.3) we have replaced C and S by respectively $C_{(s)}$ and $S_{(s)}$.

Let $\mathbb{R}^2 \setminus \{b \mid s_b = o\} = \bigcup_i X_i$, where $\bigcup_i X_i$ is an union of disjoint connected components.

Then $C_{(s)}$ and $S_{(s)}$ are given by

$$C_{(s)} = \bigoplus_i C_{(s)}|_{X_i} \qquad S_{(s)} = \bigoplus_i S_{(s)}|_{X_i}$$

where by $C_{(s)}|_{X_i}$ (or $S_{(s)}|_{X_i}$) we mean that $C_{(s)}$ (or $S_{(s)}$) is considered as a bilinear form on the subspace of functions with support in X_i.

In the same way one defines

$$K_{(s)}(x,y) = \lambda S_{(s)}(x,y)\, \Phi_{(s)}(y)$$

where $\Phi_{(s)}$ is the Gaussian variable associated with the covariance $C_{(s)}$, and one has

$$K_{(s)} = \bigoplus_i K_{(s)}|_{X_i}$$

Now supposing that the functions f_i, g_j, h_k have their supports in unit lattice squares one gets

$$S_{(s),\Lambda}(f;g;h) = \prod_i S_{(s);\Lambda\cap X_i}(f|_{X_i};g|_{X_i};h|_{X_i})$$

with obvious definitions for $f|_{X_i}$, $g|_{X_i}$ or $h|_{X_i}$.

This factorization property is one of the main input in the definition of a cluster expansion. Then one has convergence of this expansion if (see [7], [8])

1) $|Z_\Delta| > 0$ (III.1)

   where Z_Δ is the partition function in $\Lambda=\Delta$ a unit square, with Dirichlet boundary conditions on $\partial\Delta$.

2) X being one of the connected component generated by Γ, there exists K_1 large enough, a constant $C(n,N)$ and a Schwartz norm such that for f_i, g_j and h_k with support in X one has

$$|\partial^{\Gamma} S_{(s),\Lambda\cap X} (f;g;h)| \leq c(n,N) \prod_{i=1}^{n} |f_i| \prod_{j=1}^{N} |g_j||h_j| e^{-K_1|X|} \tag{III.2}$$

Condition 1) results for example from the continuity in λ. In fact, there exists λ_o, such that for $\lambda<\lambda_o$ one has

$$|Z_\Delta| > \frac{1}{2}$$

Indeed, in [5], it is proved, following an argument of E. Seiler [1], that $\lambda>o$ being fixed there exists m large enough such that $Z_\Delta > o$.

Condition 2) will be proved in part IV. More precisely we will prove that the bound (III.2) is obtained with $C(n,N) = O(1)^n$ $O(1)\sqrt{n!}$ for λ small enough and m large enough. The constant K_1 can be taken as large as we want by taking m sufficiently large. As a consequence one has

Theorem 1.
For f_i, i=1,...n, g_j, h_j, j=1,...,N in some Schwartz space

$$\lim_{\Lambda\to\infty} Z_\Lambda^{-1} S_\Lambda (f;g;h)$$

exists and satisfies all the Osterwalder-Schrader axioms. One has also an exponential clustering from which follows

Theorem 2.
There exists a Y_2 relativistic quantum field theory satisfying the Wightman axioms and possessing a mass gap.

Among the Osterwalder-Schrader axioms, the positivity axiom follows from the Feynman-Kac formula applied for finite space time cutoff.

To conclude this part we return to the definitions of the covariances C_Γ and S_Γ. For our purpose (i.e. to obtain expressions which decouple), we need only to have covariances which "behave" (see [9]) as Dirichlet covariances. Also if C_Γ is well defined as $(-\Delta_\Gamma+m^2)^{-1}$, S_Γ can be defined in many ways. A natural one can be $(\frac{1}{i}\not\partial + m)\, C_\Gamma$, and thus we need to control the local behaviour of C_Γ (remember also that $\mathcal{X}_{1/2}$ is defined with a density $F(p)^{-1/2}$ which in x-space is a pseudo differential operator). Therefore in order to avoid technical difficulties, we have chosen to proceed as in [9]. The following lemma is essential

Lemma 1

There exists a function H defined on subsets of Z^{2*} and on the squares of the lattice cover such that

1) $o \leq h(\Gamma;\Delta,\Delta') \leq 1$

$h(Z^{2*};\Delta,\Delta') = 1$ for $\Gamma\subset Z^{2*}$, Δ,Δ' unit squares

2) if $\mathbb{R}^2 \setminus (Z^{2*}\setminus\Gamma) = \bigcup_i X_i$, an union of disjoint connected components, then

$$h(\Gamma;\Delta,\Delta') = o$$

if Δ and Δ' are not in the same X_i.

3) defining

$$H(s;\Delta,\Delta') = \sum_{\Gamma\subset Z^{2*}} \prod_{b\in\Gamma} s_b \prod_{b\notin\Gamma} (1-s_b)\, h(\Gamma;\Delta,\Delta')$$

then

$$\partial^\Gamma H(s;\Delta,\Delta') = \prod_{b\in\Gamma} \frac{d}{ds_b} H(s;\Delta,\Delta')$$

is bounded by

$$e^{-m_1\, d(\Delta,\Delta',\Gamma)}\, e^{m_2\, d(\Delta,\Delta')}$$

where $d(\Delta,\Delta')$ is the distance between Δ and Δ', $d(\Delta,\Delta',\Gamma)$ is the length of the shortest path connecting Δ and Δ' and passing through each element of Γ. Finally $m_2 > m_1$ (we omit to write the dependence of h with respect to m_1 and m_2).

For the definition of such a function and the proof of the lemma see [9] and [5].

We then set

$$C_\Gamma(x,y) = \sum_{\Delta,\Delta'} h(\Gamma;\Delta,\Delta')\chi_\Delta(x)\, C_m(x,y)\, \chi_{\Delta'}(y)$$

$$S_\Gamma(x,y) = \sum_{\Delta,\Delta'} h(\Gamma;\Delta,\Delta')\chi_\Delta(x)\, S_m(x,y)\, \chi_{\Delta'}(y)$$

With these definitions it follows that in each localized graph, the original propagators are up to a constant exactly the free ones.

IV. The Cluster Expansion: The Proof

In this part a sketch of the proof of condition 2) part III, is given.

The proof follows the lines developped in $P(\Phi)_2$ theories. The differentiations generate a sum of terms which are considered as graphs. The sum of terms is bounded by using combinatoric factors. The graphs which are convergent generate factors which compensate the combinatoric ones.

Let us see now more precisely what happens for Y_2.

A differentiation $\frac{d}{ds}$ generates two kinds of terms

1) terms coming from the differentiation of the gaussian measure $d\mu_{C_{(s)}}$ of covariance $C_{(s)}$ and given by:

$$\frac{d}{ds}\, d\mu_{C_{(s)}} = \int \frac{dC}{ds}(x,y)\, \frac{\partial^2}{\partial\Phi(x)\partial\Phi(y)}\, dxdy$$

2) terms coming from the differentiation of expressions depending directly on s: for example

$$\det_{i,k} S_F\,(h_i,g_k;\Phi)\, \det_{ren}\,(1+K)$$

These two kinds of terms contain each a derived propagator: $\frac{dC}{ds}$ for the first kind, $\frac{dS}{ds}$ for the second kind.

The combinatoric associated with differentiations will therefore treat differently a derivation generating a new term from a one deriving an already derived propagator. The combinatoric factors taking into account the multiple derivations of propagators will be controlled by the decreasing factor of lemma 1, 3) part III.

We now discuss the number of new terms generated by differentiations. The control of these terms in $P(\Phi)_2$ theories comes from the fact that a differentiation of the exponent generates a polynomial in the field Φ, i.e. a finite number of monomials in the field Φ, each of bounded degree. We have therefore to exhibit a similar phenomenon. So let us look at the terms of kind 2). First consider

$$\frac{d}{ds} \det_3 (1+K) = \frac{d}{ds} [e^{\mathrm{tr}\ \ln(1+K) - \mathrm{tr}\ K + \frac{1}{2} \mathrm{tr}\ K^2}]$$

$$= \det_3 (1+K) [\mathrm{tr} \frac{K^2}{1+K} \frac{dK}{ds}]$$

$$= \det_3 (1+K) [\mathrm{tr}\ S_F\ \Phi\ S\ \Phi \frac{dS}{ds}\Phi] \qquad \text{(IV.1)}$$

where to simplify, we have written $K = S\Phi$.
It follows that

$$\frac{d}{ds} \det_{ren} (1+K) = \det_{ren} (1+K) [\mathrm{tr}\ S_F\ A_4 + A_5]$$

with $A_4 = \Phi S \Phi \frac{dS}{ds} \Phi$

Then consider the effect of $\frac{d}{ds}$ on S_F in order to control the derivative of $\det_{i,k} S_F (h_i, g_k;\phi)$

$$\frac{d}{ds} S_F = \frac{d}{ds} \frac{1}{1+K} S = - \frac{1}{1+K} \frac{dK}{ds} \frac{1}{1+K} S + \frac{1}{1+K} \frac{dS}{ds}$$

which can be rewritten as

$$\frac{d}{ds} S_F = - S_F A_4 S_F + S_F A_3 + A_2 S_F + A_1 \qquad \text{(IV.2)}$$

where as for A_4, the A_i, $i=1,2,3$, are polynomials in the field Φ and there is always a fermion propagator between two fields Φ.

We thus see that repeated applications of $\frac{d}{ds}$ generate polynomials of increasing order in S_F, i.e. in $\frac{1}{1+K}$. But as explained in McBryan's lecture what can be easily estimate are expressions of the form

$$\left(\frac{1}{1+K}\right) \overset{\otimes n}{\Lambda^n} \mathfrak{X}_{1/2} \; \det_{ren} (1+K)$$

since the antisymmetry eliminates the n lowest eigenvalues from the determinant.

We need therefore to reobtain the antisymmetry after differentiation. This is possible because of the following remark.
After a derivation on

$$\det_{i,k} S_F (h_i, g_k; \Phi) \det_{ren} (1+K)$$

collect all terms which increase the order in S_F. This gives by expanding the determinant

$$\{ [\int S_F (x,y;\Phi) A_4 (y,x) \, dx \, dy] \det_{i,k} S_F (h_i, g_k; \Phi)$$

$$+ \sum (-1)^{k+l-1} \int S_F (h_k, y; \Phi) A_4 (y,x) S_F (x, g_l; \Phi)$$

$$dxdy \times [\text{cofactor of } S_F(h_k, g_l; \Phi) \text{ in } \det_{i,k} S_F (h_i, g_k; \Phi)] \} \times$$

$$\times \det_{ren} (1+K)$$

But this is nothing else than

$$\sum \int A_4(y,x) \begin{vmatrix} S_F(x,y;\Phi) & S_F(h_1,y;\Phi) \cdots S_F(h_N,y;\Phi) \\ S_F(x,g_1;\Phi) & \\ \vdots & \{S_F(h_i,g_k;\Phi\} \\ S_F(x,g_N;\Phi) & \end{vmatrix} dxdy$$

N being the order of the matrix $\{S_F(h_i, g_k; \Phi)\}$ (i.e. of the initial determinant).

One generates in this way a determinant of one order higher.

The terms with A_5, $S_F A_3$ and $A_2 S_F$ do not change the order of the determinant.

Finally when it is a A_1 which is produced one expands the determinant with respect to it and one generates in this way determinants of one order less.

In this way, one sees that a differentiation generates a controllable number of determinants, times some polynomials of the field Φ (the $A_i, i=1,..,5$).

The same mechanism appears for $\frac{\partial}{\partial \Phi}$ - derivations. In particular one gets the analogue of formulae IV.1 and IV.2 with the A_i replaced by A_i'.

The number of terms produced after applying the above recombination rules is then bounded by the equivalent for our case of lemma 10.2 of ref. [7]. It remains therefore to bound a generic term of this expansion. It has the form

$$\int \det_{i,k} S_F(x_i, y_k; \Phi)\; w(x_1,\ldots;y_1,\ldots)\; \det_{ren}(1+K)\; dx_1..dy_1..d\mu \tag{IV.3}$$

where $w(x_1,\ldots;y_1,\ldots)$ is a polynomial in the fields Φ made with the initial (non-derived) Φ-fields and with the A_i or A_i' or derivatives with respect to s of such A_i or A_i'.

Introducing the operator $A = K+K^*+KK^*$, to bound (IV.3) one uses the following bound due to E. Seiler [1]:

$$\left|\int \det_{i,k} S_F(x_i, y_k; \Phi)\; w(x_1,..;y_1,..)\; dx_1..dy_1..\right|^2 \det_4(1+A) \leq$$

$$\leq C \, \|w\|^2_{\mathcal{H}^*_{1/2}} \; \det_4(1+A_+)$$

where A_+ is the positive part of A and

$$\|w\|^2_{\mathcal{H}^*_{1/2}} = \int |\tilde{w}(k_1,\ldots;l_1,\ldots)|^2 \, F_m^{1/2}(k_1)..\; F_m^{1/2}(l_1)..dk_1..dl_1..$$

Using now

$$|\det_3 (1+K)|^2 = \det_4 (1+A)\ e^{-1/2 \mathrm{Tr}(KK^*)^2 - 2\ \mathrm{Re}\ \mathrm{Tr}\ K^2K^* + \frac{1}{3}\ \mathrm{Tr} A^3}$$

one sees that IV.3 is bounded by

$$C\left(\int ||w||^4_{\mathcal{H}^*_{1/2}}\ d\mu\right)^{1/4}\left(\int |\det_4(1+A_+) e^{-1/2 \mathrm{Tr}(KK^*)^2 - 2 \mathrm{Re} \mathrm{Tr} K^2 K^*} \times\right.$$

$$\left.\times\ e^{1/2 \mathrm{Tr}\ A^3}\ B_M\ B_M^*|^2 d\mu\right)^{1/2}$$

The second bracket is bounded by

Lemma 2
For any $p>1$, there exists M^2 large enough such that

$$||\det_{ren}\ (1+K_1)||_p \le ||\det_4(1+A_+)\ e^{-1/2\ \mathrm{Tr}\ (KK^*)^2} \times$$

$$\times\ e^{-2\ \mathrm{Re}\ \mathrm{Tr}\ K^2K^* + 1/3\ \mathrm{Tr}\ A^3}\ B_M B_M^*||^{1/2}_{P/2} \le e^{K_2|\Lambda|}$$

for some $K_2(p)$ independent of Λ.

This lemma will be shortly discussed in part V.

The first bracket is estimated using the method of localized graphs as in [6] and [9]. Taking m large enough with respect to m_2 one uses the fact it is possible to extract decreasing factor $e^{-md(x,y)}$ from each propagator to control the combinatoric factor. Using also the convergence of the graphs one can extract a factor $m^{-\varepsilon}$ for each derived propagator and this factor ensures the overall convergence of the cluster expansion.

V. Volume Dependence of the Upper Bound

This part is devoted to the proof of lemma 2. The method we follow to prove the lemma is very close in spirit with the technique developed by Seiler [1] in proving the existence of a cutoff theory. Since McBryan has explained in his lecture how to get such a bound, I will only insist here on the difference with his method.

First we write

$$\det_4(1+A_+)\; e^{-1/2\ \mathrm{Tr}(K\,K^*)^2}\; e^{-2\ \mathrm{Re}\ \mathrm{Tr}K^2K^* +\ 1/3\ \mathrm{Tr}\ A^3} B_M B_M^*$$

as $L\, e^T$ where

$$L = \det_4(1+A_+)\; e^{-1/2 :\mathrm{Tr}(K+K^*)^2: -1/2\ \mathrm{Tr}\ (KK^*)^2 - 2\ \mathrm{Re}\ \mathrm{Tr}K^2K^*}$$

$$\times\; e^{1/3\ \mathrm{Tr}\ A^3}$$

$$T = \frac{1}{2} : \mathrm{Tr}\ (K+K^*)^2: - \mathrm{Tr}_{\mathrm{reg}} :K^2: - M^2 \int :\Phi^2:(x)\ dx$$

$$= \mathrm{Tr}_{\mathrm{reg}} : KK^*: - M^2 \int :\Phi^2:(x)\ dx$$

(for the definition of $\mathrm{Tr}_{\mathrm{reg}}:K^2:$ or of $\mathrm{Tr}_{\mathrm{reg}}:KK^*:$ see [1],[2] or [5])

Now it is easy to see that

$$||\mathrm{Tr}_{\mathrm{reg}} : K\,K^* :||_{H.S.} < \infty$$

from which it follows that $p>1$, being given, there exists $M^2(p)$ such that for $M^2 \geq M^2(p)$ one has

$$e^T \in L_p\ (d\mu)$$

and the norm is bounded by $e^{C|\Lambda|}$ for some constant C.

It remains therefore to bound L.

We introduce a sequence of cutoff ρ_n, $n \in Z^*$ and $\rho_n \to \infty$ for $n \to \infty$. Then to each lattice square Δ we assign a cutoff K_Δ taken in this sequence and correspondingly we define cutoff fields Φ_K by

$$\Phi_K(x) = \sum_\Delta \chi_\Delta(x) \int \tilde{\Phi}(k)\, \eta_{K_\Delta}(k)\, dk$$

where η_k is some cutoff function, and cutoff operators K_K and A_K.

The idea of Seiler is then to compare the <u>exponent</u> of L with the exponent of L_K obtained by replacing Φ by Φ_K

Proceeding in this way one gets that

$$L \leq e^{O(1)\sum_\Delta\{(\ln K_\Delta)^3 + V_\Delta(K_\Delta)\}}$$

where $V_\Delta(K_\Delta)$ is some polynomial in the field Φ depending only on the cutoff K_Δ which has the property that:

$$\int |V_\Delta(K_\Delta)|^2\, d\mu \leq O(1)\, K_\Delta^{-\varepsilon} \qquad \text{(V.1)}$$

for some $\epsilon > 0$.

One then defines in path space

$$\mathcal{D}_1(\Delta) = \text{set of points where } V_\Delta(\rho_1) \leq 1$$

$$\vdots$$

$$\mathcal{D}_n(\Delta) = \text{set of points where } V_\Delta(\rho_{n-1}) > 1 \text{ and } V_\Delta(\rho_n) < 1$$

thus $\bigcup_{i=1}^{n} \mathcal{D}_i(\Delta)$ = set of points where $V_\Delta(\rho_n) < 1$

and since from (V.1) (see [10]) the measure of the set of points where $V_\Delta(\rho_n) \geq 1$ tends to zero with ρ_n, $\{\mathcal{D}_i(\Delta)\}_{i \in Z^+}$ defines a partition of the path space.

We can therefore proceed as in Dimock and Glimm [11] and we get that for some constant C'

$$\int L^P \, d\mu \le e^{C'|\Lambda|}$$

Finally let us remark that the finite renormalization of the boson mass is probably unnecessary and that one can proceed as in [12] and [13].

VI. Conclusions

The result presented here is only a preliminary step in the study of the Y_2 model. Many interesting questions have to be solved which can go far beyond this existence proof. First it will be interesting to extend to this model the results obtained for $P(\Phi)_2$ theories. For instance, extension of the theory to complex values of the coupling constant and question of the summability properties. Can we prove the existence of phase transitions for Y_2? Secondly among all questions typically related to this model, one is of particular interest. Can we define some zero mass limit for the bose field although dimension two is very pathological for such kind of question?

References

[1] E. SEILER, Schwinger Functions for the Yukawa model in two dimensions with space time cut off, Comm. Math. Phys., 42, 2 (1975)

[2] O. McBRYAN, Volume dependence of Schwinger Functions in the Yukawa $_2$ Quantum Field Theory, Rockefeller University preprint.

[3] E. SEILER, B. SIMON, Bounds in the Yukawa$_2$ Quantum Field Theory: Upper Bound on the Pressure, Hamiltonian Bound and Linear Lower Bound, Princeton University preprint.

[4] O. McBRYAN, Convergence of the Vacuum Energy Density, Φ-bounds and Existence of Wightman Functions for the Yukawa$_2$ model, Rockefeller University preprint

[5] J. MAGNEN, R. SENEOR, The Wightman Axioms for the Weakly Coupled Yukawa Model in two Dimensions, E. Polytechnique, Palaiseau, preprint.

[6] J. GLIMM, A. JAFFE, Positivity of the Φ_3^4 Hamiltonian, Forts. der Phys., 21, 327-376 (1973).

[7] J. GLIMM, A. JAFFE, T. SPENCER, The Cluster Expansion, in Constructive Quantum Field Theory, Lecture notes in Physics, no. 25, Springer, Berlin (1973).

[8] J.P. ECKMANN, J. MAGNEN, R. SENEOR, Decay Properties and Borel summability for the Schwinger Functions in $P(\Phi)_2$ theories, Comm. Math. Phys., 39, 4 (1975)

[9] J. MAGNEN, R. SENEOR, The Infinite Volume Limit of the Φ_3^4 model, to appear in Ann. Inst. H. Poincaré.

[10] E. NELSON, Probability Theory and Euclidean Field Theory, in Constructive Qunatum Field Theory, Lecture Notes in Physics, No. 25, Springer, Berlin (1973).

[11] J. DIMOCK, J. GLIMM, Measures on the Schwartz distribution space and Application to $P(\Phi)_2$ field theories, to appear in Adv. Math.

[12] O. McBRYAN, Finite mass renormalizations in the Yukawa_2 Quantum Field Theory, Comm. Math. Phys. to appear.

[13] E. SEILER, B. SIMON, On finite mass renormalizations in the two dimensional Yukawa model, Princeton University preprint.

For a more complete list of references, see [2], [3] or [5].

Acta Physica Austriaca, Suppl. XVI, 125–146 (1976)

External Field Dependence of Magnetization and Long Range Order in Quantum Field Theory*+

Francesco Guerra**
Institute for Advanced Study, Princeton,
New Jersey o854o

* Expanded version of a talk given at the Symposium "Mathematical Problems of Quantum Dynamics", held at ZiF, Bielefeld, September 8 - 12.

+ Research sponsored in part by the National Science Foundation Grant No. GP-4o768X.

** Address after April 3o, 1976: Via Aniello Falcone 7o, 8o127 Napoli. Italy

1. Introduction

We consider field theoretical models of self-interacting bosons in the framework of the Euclidean formulation of quantum field theory advanced by Symanzik [26] and Nelson [18]. We employ methods of classical statistical mechanics according to the program advocated in [13], see also [22] and [14].

For the sake of simplicity we take mainly into account the $P(\phi)_2$ model of a Bose field with polynomial self-interaction in two-dimensional space-time [27,22], but our consideration extend also to more general ferromagnetic lattice systems [12] in an arbitrary number of dimensions or to the ϕ_3^4[8] quantum field theory for which there has been great progress recently both in the weak and strong coupling case [2,17,3,21,5,19]. Our methods extend also to the ϕ_4^4 theory, provided the relevant problems of renormalization and existence can be solved.

All these systems are described through a local order parameter $\phi(x)$. We let the self-coupled system interact with a uniform external field of strength λ, coupled linearly with the order parameter ϕ, and study the dependence of the free energy density and other properties of the infinite volume equilibrium states in function of λ. Our main results refer to the magnetization $M(\lambda) = \langle\phi(x)\rangle$ and the long range order $C^2 = \lim_{|x-y|\to\infty} \langle\phi(x)\phi(y)\rangle_T$. For a large class of states we prove that the magnetization equals the derivative of the free energy and the long range order vanish for all values of the λ for which the free energy is differentiable.

The organization of the report is as follows. In Section 2 we recall all relevant properties of the free energy as a function of the external field λ. In Section 3 we introduce a class of Euclidean invariant states satisfy inequalities of F K G type

[4,13] and a regularity assumption. For these states we prove various inequalities involving the derivatives of the free energy, the magnetization, the long range order and the susceptibility (Section 4). In Section 5 we consider the states obtained by letting the external field increase or decrease to a definite value and study their properties. Section 6 is devoted to the particular case of states obtained through Nelson monotone convergence theorem [18] from finite volume approximations with zero field boundary conditions for even self-interactions.

In conclusion the author would like to thank the Organizing Committee of the Symposium for the kind hospitality extended to him at ZiF in Bielefeld.
Very useful conversations with Jürg Fröhlich, Lon Rosen and Barry Simon are also gratefully acknowledged.

2. The Free Energy Density

First of all let us consider the case of $P(\phi)_2$. For the real semi-bounded interaction polynomial P we write $P(\phi) = Q(\phi) - \lambda\phi$ where Q contains all terms starting from the second order. We introduce also the self-interaction energy for a bounded region $\Lambda \subset \mathbb{R}^2$

$$U_\Lambda = \int_\Lambda :Q(\phi(x)) : dx$$

and the partition function

$$Z^X_\Lambda = \langle \exp(-U_\Lambda + \lambda\phi(\chi_\Lambda)) \rangle^X_\Lambda$$

where χ_Λ is the characteristic function of Λ and $\langle\rangle^X_\Lambda$ denotes the expectation for the free Euclidean-Markov field in the region Λ with given boundary conditions X. For a large class of boundary conditions X, the results of [11,13,14] show the existence of the limit

$$\alpha_\infty = \lim_{\Lambda \to \mathbb{R}^2} |\Lambda|^{-1} \log Z^X_\Lambda$$

and its independence from the boundary conditions X. For the free energy, or pressure, α_∞ we write $\alpha_\infty(Q,\lambda)$ to emphasize the dependence on Q and λ or $\alpha_\infty(\lambda)$ when Q is fixed.

Proposition 1. The free energy $\alpha_\infty(\lambda)$ is a convex function of λ and therefore is also continuous. Moreover there are positive constants K_1 and K_2 such that for large $|\lambda|$ we have

$$K_1|\lambda| \leq \alpha_\infty(\lambda) \leq K_2\lambda^2$$

Proof. The convexity of $\alpha_\infty(Q,\lambda)$ in Q and λ is an elementary consequence of Hölder's inequality. We have therefore

$$2\alpha_\infty(Q,\lambda) \leq \alpha_\infty(2Q,0) + \alpha_\infty(0,2\lambda)$$

By direct calculation we have $\alpha_\infty(0,\lambda) = \lambda^2/2m_0^2$, where m_0 is the bare mass, and the upper bound is established. If we write $Q = \overline{Q} + Q'$, where $\overline{Q}$ and Q' are the even and odd parts of Q, then $\alpha_\infty(\overline{Q} - Q', 0) + \alpha_\infty(\overline{Q}+Q',\lambda) \geq 2\alpha_\infty(\overline{Q}, \lambda/2)$. But it is easy to see that for large $|\lambda|$ $\alpha_\infty(\overline{Q}, \lambda/2) \geq K|\lambda|$, with $K > 0$, because the magnetization in finite volume is positive for $\lambda > 0$. The lower bound follows.

Let λ_0 be the value of λ for which $\alpha_\infty(\lambda)$ takes its minimum value. Then $\alpha_\infty(\lambda)$ is increasing in λ for $\lambda \geq 0$ and decreasing for $\lambda \leq 0$. If Q is even then obviously $\lambda_0 = 0$.

Standard properties of convex functions give

Proposition 2. The left and right derivatives

$$M^{(\pm)}(\lambda) = \pm\lim_{\varepsilon\to0^+} \varepsilon^{-1}[\alpha_\infty(\lambda\pm\varepsilon)-\alpha_\infty(\lambda)]$$

exist for any λ and they are equal for almost all values of λ. The following properties hold

a) $M^{(-)}(\lambda) \leq M^{(+)}(\lambda) \leq M^{(-)}(\lambda')$ for $\lambda < \lambda'$

b) $M^{(\pm)}(\lambda)$ are increasing in λ and

$$M^{(\pm)}(\lambda) \leq 0 \text{ for } \lambda < \lambda_0,\ M^{(\pm)}(\lambda) \geq 0 \text{ for } \lambda > \lambda_0$$

$$M^{(+)}(\lambda_0) \geq 0,\ M^{(-)}(\lambda_0) \leq 0$$

c) $M^{(+)}(\lambda)$ is upper semicontinuous and $M^{(-)}(\lambda)$ is lower semicontinuous.

If the coefficients of Q and λ are small enough then $\alpha_\infty(\lambda)$ is analytic in λ and obviously $M^{(+)}(\lambda) = M^{(-)}(\lambda)$, as a consequence of the convergence of the Glimm-Jaffe-Spencer cluster expansion [9]. By an argument of Spencer [25] the same results hold for any Q if $|\lambda|$ is large enough.

When Q is even then $\alpha_\infty(\lambda) = \alpha_\infty(-\lambda)$ and $M^{(+)}(\lambda) = -M^{(-)}(-\lambda)$. If Q is a fourth order polynomial then by the classical Ising approximation of Simon and Griffiths [24] the Lee-Yang theorem [20] and the Griffiths-Hurst-Sherman inequality [10] holds. As a consequence $\alpha_\infty(\lambda)$ is analytic for Re $\lambda \neq 0$, moreover $M^{(+)}(\lambda) = M^{(-)}(\lambda)$ for Re $\lambda \neq 0$ and $M^{(\pm)}(\lambda)$ is concave, therefore continuous, for $\lambda > 0$.

All properties stated in this section for the dependence of α_∞ on the strength of the external field λ hold also for lattice systems like those considered in [12] and for their limits when the lattice spacing goes to zero.

3. States with Thermodynamic Stability

The inequalities of Fortuin-Kastelejn-Ginibre type, established in [4] and extended to Euclidean quantum field theory in [13], will play a relevant role in the following. Here we recall some basic facts.

Definition 3. A numerical function $F : \mathbb{R}^n \to \mathbb{R}$ is called increasing if $F(x) \leq F(x')$ for $x_i \leq x_i'$, $i = 1,\ldots n$. A function F of the fields $\phi(x)$ is called increasing if $F = F(\phi(f_1),\ldots,\phi(f_n))$ for some increasing numerical function F and test functions $f_i \geq 0$. Examples of increasing functions of the fields are [22]: $\phi(f)$, $\sigma(f)$, $\phi(f) - \sigma(f)$, $\Pi_i\rho(f_i)$, $\Sigma_i\rho(f_i)$, $\Sigma_i\rho(f_i) - \Pi_i\rho(f_i)$, $\Sigma_i\phi(f_i) - \Pi_i\rho(f_i)$, where f, $f_i \geq 0$ and

$$\sigma(f) = \begin{cases} \phi(f) \text{ if } |\phi(f)| \leq 1 \\ \operatorname{sgn} \phi\ (f) \text{ if } |\phi(f)| > 1 \end{cases}$$

$$\rho(f) = \frac{1}{2}(1 + \sigma(f))$$

Definition 4. A state $\langle\ \rangle$ is called FKG state if any two increasing functions F, G of the fields are positively correlated, i.e.

$$\langle FG\rangle \geq \langle F\rangle \langle G\rangle$$

(whenever the averages have meaning). From the definition of FKG state it follows immediately

Proposition 5. If for a given state $\langle\ \rangle$ and some fixed increasing G the states defined by

$$\langle\cdot\rangle_s = \langle\cdot\, e^{sG}\rangle/\langle e^{sG}\rangle$$

are FKG for s in some interval $[s_1, s_2]$ then the average $\langle F\rangle_s$ is increasing in s for all increasing functions F of the fields. We also have

Proposition 6. If $\langle\ \rangle$ and $\langle\ \rangle'$ are FKG states and $\langle\ \rangle$ is dominated by $\langle\ \rangle'$ in the FKG sense, i.e. $\langle F\rangle \leq \langle F\rangle'$ for any increasing F, then any convex linear combination of $\langle\ \rangle$ and $\langle\ \rangle'$ is FKG.

Proof. Let $\langle\ \rangle_t = t\langle\ \rangle + (1-t)\langle\ \rangle'$, $0 \leq t \leq 1$. Then for increasing F and G we have

$$\langle FG\rangle_t = t\langle FG\rangle + (1-t)\langle FG\rangle' \geq t\langle F\rangle\langle G\rangle + (1-t)\langle F\rangle'\langle G\rangle' =$$

$$= \langle F\rangle_t\langle G\rangle_t + t(1-t)(\langle F\rangle' - \langle F\rangle)(\langle G\rangle' - \langle G\rangle) \geq \langle F\rangle_t\langle G\rangle_t$$

For a given interaction $Q(\phi) - \lambda\phi$, with Q fixed and $\lambda \in \mathbb{R}$, we consider a family of Euclidean invariant FKG states $\langle\ \rangle_\lambda$ obtained through some definite limiting procedure from volume cut off approximations with given boundary conditions. For example we can consider weak coupling states arising from the Glimm-Jaffe-Spencer cluster expansion [9], or states in large external field like in the construction given by Spencer [25], or states obtained through the infinite volume limit of local linear perturbations to Spencer states according to the procedure of Fröhlich and Simon [6]. If Q is even we can consider the Dirichlet states obtained from Nelson monotonicity theorem [18,13], see Section 6. More general types of states can be considered, like those arising from compactness arguments analogous to the construction carried on by Glimm and Jaffe in [7].

In any case for the states $\langle\ \rangle_\lambda$ we make the following assumption

Thermodynamic stability. For any rectangle Λ the following inequality holds

$$|\Lambda|^{-1}\log\langle\exp(s\phi(\chi_\Lambda))\rangle_\lambda \leq \alpha_\infty(\lambda+s) - \alpha_\infty(\lambda)$$

Then we have immediately

Theorem 7. As $\Lambda \to \infty$ (Fisher)

$$\lim|\Lambda|^{-1}\log\ \exp(s\phi(\chi_\Lambda))\rangle =$$

$$= \sup{}' |\Lambda|^{-1} \log\langle \exp(s\phi(\chi_\Lambda))\rangle \leq \alpha_\infty(\lambda+s) - \alpha_\infty(\lambda)$$

where sup' means the supremum for all rectangles Λ.

Proof. Follows immediately from FKG because $\pm \exp(s\phi(\chi_\Lambda))$ are increasing functions if $s \geq 0$ or $s \leq 0$ respectively. Therefore for any s $\langle\exp(s\phi(\chi_\Lambda))\rangle \geq \langle\exp(s\phi(\chi_{\Lambda_1}))\rangle\langle \exp(s\phi(\chi_{\Lambda_2}))\rangle$, if Λ_1 and Λ_2 are disjoint regions with union Λ. The superadditivity of log $\langle\ \rangle$ gives the theorem by well known arguments, see for example [20].

We would like to point out the relation between thermodynamical stability and the asymptotic behaviour of the solutions of Dobrushin-Lanford-Ruelle equations [1,15] extended to $P(\phi)_2$ in [13].

If F is a function of the fields in the bounded region Λ then, for a given interaction $Q(\phi)-\lambda\phi$, we have

$$\langle F\rangle_\lambda = Z_\Lambda^{-1}\langle F \exp\ (-U_\Lambda+\lambda\phi(\chi_\Lambda))\psi_{\partial\Lambda}\rangle_{(0)}$$

where $\langle\ \rangle_{(0)}$ are the expectations for the free Euclidean-Markov field with free boundary conditions, $U_\Lambda = \int_\Lambda :Q(\phi(x)):dx$, $Z_\Lambda = \langle\exp(-U_\Lambda+\lambda\phi(\chi_\Lambda))\rangle_{(0)}$, and $\psi_{\partial\Lambda}$ is a boundary correction function of the fields at the boundary.

In [6] Fröhlich and Simon obtain a good control of the solutions of DLR equations in $P(\phi)_2$. From this analysis it follows in particular that for large circles Λ there are constants c_p and η with $\frac{1}{2} < \eta < 1$ such that

$$||\psi_{\partial\Lambda}||_p \leq \exp|c_p|\Lambda|^\eta|$$

for any p, $1 \leq p < \infty$.

Through a straightforward application of Hölder's inequality we have immediately

$$\overline{\lim}|\Lambda|^{-1}\log \langle \exp(s\phi(\chi_\Lambda))\rangle \leq p^{-1}\alpha_\infty(pQ,p(\lambda+s)) - \alpha_\infty(Q,\lambda)$$

for any $p>1$. Therefore taking the limit $p \to 1$ we establish thermodynamic stability as a consequence of the tempered behavior of the DLR correction $\psi_{\partial\Lambda}$ as $\Lambda \to \infty$.

We believe that for all reasonable states the limit proven in Theorem 7 is in fact equal to $\alpha_\infty(\lambda+s) - \alpha_\infty(\lambda)$, see for example Section 6. Through a very simple argument based on conditioning theory between Dirichlet and free states [13,14] and Jensen inequality, this can be proven under the assumption, so far not established yet, that

$$\lim|\Lambda|^{-1}<\log\psi_{\partial\Lambda}>_{(0)} = 0$$

In fact we have

$$<\exp(s\phi(\chi_\Lambda))>_\lambda = Z_\Lambda^{-1}<\exp(-U_\Lambda+(\lambda+s)\phi(\chi_\Lambda))\psi_{\partial\Lambda}>_{(0)} \geq$$

$$\geq Z_\Lambda^D(Q,\lambda+s)Z_\Lambda^{-1}(Q,\lambda)\exp(-<\log\psi_{\partial\Lambda}>_{(0)})$$

which establishes equality in the limit of Theorem 7 under the stated assumption.

4. Magnetization and Long Range Order

Since our states are translation invariant we define the magnetization $M(\lambda)$ through

$<\phi(f)>_\lambda = M(\lambda)\int f(x)dx,$ or formally

$M(\lambda) = <\phi(x)>_\lambda.$ Then we have

Theorem 8. The following inequalities hold

$$M^{(-)}(\lambda) \leq M(\lambda) \leq M^{(+)}(\lambda)$$

Therefore $M(\lambda)$ is increasing in λ and $M(\lambda\pm) = M^{(\pm)}(\lambda)$. Moreover for all values of λ for which $M^{(+)}(\lambda) = M^{(-)}(\lambda)$ we have $M(\lambda) = M^{(\pm)}(\lambda)$.

Proof. By Jensen inequality

$$<\exp(s\phi(\chi_\Lambda))>_\lambda \geq \exp(s<\phi(\chi_\Lambda)>_\lambda = \exp s|\Lambda|M(\lambda)$$

But by thermodynamic stability we have

$$s\,M(\lambda) \leq \alpha_\infty(\lambda+s) - \alpha_\infty(\lambda)$$

and taking $s \to 0^+$ or $s \to 0^-$ we get the inequalities. The rest follows easily from Proposition 2.

Let us consider now the two-point function

$$S(x-y)=<\phi(x)\phi(y)>_\lambda=C^2(\lambda)+M^2(\lambda)+\int_0^\infty \rho(m^2)S_{m^2}(x-y)dm^2$$

where S_{m^2} is the free Schwinger function for mass m and $C^2(\lambda)$ defines the long range order. As a consequence of FKG inequalitie

Simon [22] has proven that the vanishing of $C^2(\lambda)$ implies the cluster property for the state $< >_\lambda$. Therefore $< >_\lambda$ is pure if and only if $C^2(\lambda) = 0$.

In order to establish inequalities on $C^2(\lambda)$, first of all let us state

Theorem 9. As $\Lambda \to \infty$ (Fisher)

$$\lim|\Lambda|^{-2}<\phi^2(\chi_\Lambda)>_\lambda=\inf{}'|\Lambda|^{-2}<\phi^2(\chi_\Lambda)>_\lambda=C^2(\lambda)+M^2(\lambda)$$

where inf' denotes the infimum over all rectangles Λ.

Proof. The convergence follows from the inequality

$$||\phi(\chi_\Lambda)||_2 \leq ||\phi(\chi_{\Lambda_1})||_2 + ||\phi(\chi_{\Lambda_2})||_2$$

for Λ_1,Λ_2 disjoint regions with union Λ and $||\phi(\chi_\Lambda)||_2^2=<\phi^2(\chi_\Lambda)>_\lambda$ and standard subadditivity arguments. On the other hand, since $|\Lambda|^{-2}S_{m^2}(\chi_\Lambda,\chi_\Lambda)$ tends to zero for all values of $m^2> 0$, the spectral part of S does not give any contribution in the limit by the dominated convergence theorem (take $\Lambda \to \infty$ through rectangles).

Remark. By the same reasoning we have also convergence for all expressions

$$|\Lambda|^{-1}<|\phi(\chi_\Lambda)|^p>^{\frac{1}{p}} \to D_p(\lambda) \quad , \; 1 \leq p < \infty$$

and

$$D_p(\lambda) \leq D_{p'}(\lambda) \qquad \text{for } p \leq p'$$

Lemma 10. Let ϕ be a random variable for which $<\exp(\pm\phi)< \infty$, then

$$<\cosh\phi> \geq \cosh <\phi^2>^{\frac{1}{2}}$$

Proof. $<\cosh\phi> = \sum_{n=0}^{\infty} <\phi^{2n}>/(2n)! \geq \sum_{n=0}^{\infty} <\phi^2>^n/(2n)! = \cosh <\phi^2>^{\frac{1}{2}}$.
Now we state the main result of this section.

Theorem 11. The following inequality holds

$$C^2(\lambda) \leq (M^{(+)}(\lambda) - M(\lambda))(M(\lambda) - M^{(-)}(\lambda))$$

Proof. Define $\bar{M}$ so that

$$\alpha_\infty(\lambda+s) - \alpha_\infty(\lambda) - s\bar{M} = s\bar{M} + \alpha_\infty(\lambda-s) - \alpha_\infty(\lambda)$$

Putting $\phi = s(\phi(\chi_\Lambda) - |\Lambda|\bar{M})$ in Lemma 10 and using $\cosh x \geq \frac{1}{2}e^x$ and thermodynamic stability we have

$$\exp(s<(\phi(\chi_\Lambda) - |\Lambda|\bar{M})^2>) \leq 2\exp|\Lambda|(\alpha_\infty(\lambda+s) - \alpha_\infty(\lambda) - s\bar{M})$$

If we take the logarithm, divide by $|\Lambda|$, let $|\Lambda| \to \infty$ and then $s \to 0^+$, we have immediately the theorem, because $\bar{M}\to\frac{1}{2}(M^{(+)}(\lambda) - M^{(-)}(\lambda))$ as $s \to 0^+$.

Corollary 12. If $\bar{M}(\lambda) = M^{(+)}(\lambda)$ or $M(\lambda) = M^{(-)}(\lambda)$ then $< >_\lambda$ is a pure phase, i.e. $C^2(\lambda) = 0$. In particular $< >_\lambda$ is pure for all values of λ for which the free energy $\alpha_\infty(\lambda)$ is differentiable, i.e. $M^{(+)}(\lambda) = M^{(-)}(\lambda)$.

Finally we discuss the existence of the susceptibility $\chi(\lambda)$. Introduce the truncated two point function

$$<\phi(f)\phi(g)>_T = <\phi(f)\phi(g)> - <\phi(f)><\phi(g)>$$

then we have

Theorem 13. For disjoint regions Λ_1,Λ_2 with union Λ

$$<\phi^2(\chi_\Lambda)>_T \geq <\phi^2(\chi_{\Lambda_1})>_T + <\phi^2(\chi_{\Lambda_2})>_T$$

Therefore as $\Lambda \to \infty$ (Fisher)

$$\lim|\Lambda|^{-1}<\phi^2(\chi_\Lambda)>_T = \sup{}'|\Lambda|^{-1}<\phi^2(\chi_\Lambda)>_T \equiv \chi(\lambda)$$

Proof. We have $<\phi^2(\chi_\Lambda)> = <\phi^2(\chi_{\Lambda_1})>+<\phi^2(\chi_{\Lambda_2})>+2<\phi(\chi_{\Lambda_1})\phi(\chi_{\Lambda_2})>$ but by FKG $<\phi(\chi_{\Lambda_1})\phi(\chi_{\Lambda_2})> \geq <\phi(\chi_{\Lambda_1})><\phi(\chi_{\Lambda_2})> = M^2(\lambda)|\Lambda_1||\Lambda_2|$. Since $|\Lambda| = |\Lambda_1|+|\Lambda_2|$, superadditivity follows and the rest is standard. Theorem 13 defines the susceptibility $\chi(\lambda)$. For particular values of λ it could be $\chi(\lambda) = \infty$, but we have

<u>Theorem 14.</u> If $\chi(\lambda) < \infty$ then $C^2(\lambda) = 0$.

<u>Proof.</u> If $\lambda < \infty$ then $|\Lambda|^{-2}\langle\phi^2(\chi_\Lambda)\rangle_\lambda \leq M^2(\lambda) + |\Lambda|^{-1}\chi(\lambda)$, for Λ rectangle, and taking $\Lambda \to \infty$ we have $C^2(\lambda) = 0$ as a consequence of Theorem 9.

<u>Theorem 15.</u> If $C^2(\lambda) = 0$ and the physical mass m_{ph} is positive, then $\chi(\lambda) \leq m_{ph}^{-2}$.

<u>Proof.</u> Let us recall that the physical mass m_{ph}^2 is the bottom of the support of ρ in the spectral representation of the two point function. But we have $S_{m^2}(\chi_\Lambda, \chi_\Lambda) \leq S_{m^2_{ph}}(\chi_\Lambda, \chi_\Lambda)$,

$$\lim |\Lambda|^{-1} S_{m^2_{ph}}(\chi_\Lambda, \chi_\Lambda) = m_{ph}^{-2} \quad \text{and} \quad \int \rho(m^2)\, dm^2 = 1,$$

and the theorem follows.

5. The States $< \ >_\lambda^{(\pm)}$

For any increasing function F of the fields the expectation value $\langle F\rangle_\lambda$ is an increasing function of λ as a consequence of the FKG inequalities, therefore we can define the states $< \ >_\lambda^{(\pm)}$ in the following way

$$\langle F\rangle_\lambda^{(+)} = \inf_{s>0} \langle F\rangle_{\lambda+s} \ , \quad \langle F\rangle_\lambda^{(-)} = \sup_{s>0} \langle F\rangle_{\lambda-s}$$

Obviously we have

$$\langle F\rangle_\lambda^{(-)} \leq \langle F\rangle_\lambda \leq \langle F\rangle_\lambda^{(+)}$$

and by Proposition 6 any convex linear combination of $< \ >_\lambda^{(\pm)}$ is FKG.

For $f \geq 0$ $\phi(f)$ is increasing therefore

$$\langle\phi(f)\rangle_\lambda^{(+)} = \inf_{s>0} \langle\phi(f)\rangle_{\lambda+s} = \int f dx \inf_{s>0} M(\lambda+s) = \int f dx\, M^{(+)}(\lambda)$$

Since an analogous result holds for $\langle\phi(f)\rangle_\lambda^{(-)}$ we have

<u>Theorem 16.</u> $\langle\phi(f)\rangle_\lambda^{(\pm)} = \int f\, dx\, M^{(\pm)}(\lambda)$ respectively. If $M^{(+)}(\lambda) = M^{(-)}(\lambda)$ then $\langle\phi(f)\rangle_\lambda^{(\pm)} = \langle\phi(f)\rangle_\lambda = M(\lambda)\int f dx = M^{(\pm)}(\lambda)\int f\, dx$.

The interest of this theorem is in the fact that it shows that the one point function for the states $< \ >_\lambda^{(\pm)}$ does not depend on the boundary conditions entering in the construction of $< \ >_\lambda$ but only on the interaction. We have also

<u>Theorem 17.</u> If $M(\lambda) = M^{(\pm)}(\lambda)$ then $< \ >_\lambda = < \ >_\lambda^{(\pm)}$ respectively. If $M^{(-)}(\lambda) = M^{(+)}(\lambda)$ then $< \ >_\lambda = < \ >_\lambda^{(\pm)}$, i.e. the continuity of the one point function implies the continuity of all correlation functions.

Proof. Consider the first statement. For $f_i \geq 0$, according to Definition 3 the functions $\rho \equiv \Pi_i \rho(f_i)$ and $\Sigma_i \phi(f_i) - \rho$ are increasing, therefore

$$0 \leq \langle\rho\rangle_\lambda^{(+)} - \langle\rho\rangle_\lambda \leq \Sigma_i (\langle\phi(f_i)\rangle_\lambda^{(+)} - \langle\phi(f_i)\rangle_\lambda) =$$

$$= \Sigma_i \int f_i(x)\, dx\, (M^{(+)}(\lambda) - M(\lambda))$$

and the theorem follows. The rest is analogous.

Remark. The use of FKG inequalities in the proof of Theorem 17 has been suggested by analogous methods employed by Lebowitz and Presutti [16] in their study of general continuous spin systems, in order to show the independence of states from boundary conditions in some cases.

Theorem 18. If $\langle\ \rangle_\lambda$ have thermodynamic stability the same happens for $\langle\ \rangle_\lambda^{(\pm)}$.

Proof. Since $\pm \exp(s\phi(\chi_\Lambda))$ are increasing respectively for $s \geq 0$ and $s \leq 0$, the proof is a trivial consequence of the inequalities

$$\langle F\rangle_\lambda^{(-)} \leq \langle F\rangle_\lambda \leq \langle F\rangle_\lambda^{(+)} \qquad \text{and}$$

$$\langle F\rangle_{\lambda'}^{(+)} \leq \langle F\rangle_\lambda \leq \langle F\rangle_{\lambda''}^{(-)} \qquad \text{for } \lambda' < \lambda < \lambda''$$

and the continuity of $\alpha_\infty(\lambda)$ in λ.

An immediate consequence of Theorems 11 and 16 is $C^2_{(\pm)}(\lambda) = 0$. Therefore we have

Theorem 19. The states $\langle\ \rangle_\lambda^{(\pm)}$ are pure.

It would be very nice to prove that the states $\langle\ \rangle_\lambda^{(\pm)}$ do not depend on the boundary conditions but only on the interaction and that any state $\langle\ \rangle_\lambda$ is a convex combination of $\langle\ \rangle_\lambda^{(\pm)}$.

Finally we have

<u>Theorem 20.</u> If F is increasing then $\langle F\rangle_{\lambda}^{(\pm)}$ are respectively upper and lower semicontinuous in λ. Therefore $\langle F\rangle_{\lambda+}^{(\pm)} = \langle F\rangle_{\lambda}^{(+)}$ and $\langle F\rangle_{\lambda-}^{(\pm)} = \langle F\rangle_{\lambda}^{(-)}$.

<u>Proof.</u> Since $\langle F\rangle_{\lambda}$ is almost everywhere continuous in λ, in the definition of $\langle F\rangle_{\lambda}^{(\pm)}$ we can choose sequences s_n such that $s_n \to 0$ and $\lambda \pm s_n$ are points of continuity, then the theorem follows.

6. Dirichlet States

In this section we assume Q even and take as states $< >_\lambda$ those arising from Nelson monotone convergence theorem [18,13]. Then both FKG and GKS inequalities hold [13]. We assume $\lambda \geq 0$ without loss of generality.
First of all let us prove the thermodynamical stability.

Theorem 21. For $s \in \mathbb{R}$, as $\Lambda \to \infty$ (Fisher)

$$\lim |\Lambda|^{-1} \log < \exp(s\phi(\chi_\Lambda))> = \sup{}' |\Lambda|^{-1} \log < \exp(s\phi(\chi_\Lambda)) > \leq$$

$$\leq \alpha_\infty(\lambda+s) - \alpha_\infty(\lambda)$$

where the supremum is over all rectangles. For $s \geq 0$ the equality holds.

Proof. We call $< >_\Lambda$ the volume cut off half-Dirichlet state in Λ and $< >^P_\Lambda$ the state with periodic boundary conditions, if Λ is a rectangle. Assume $s \geq 0$ and let Λ and Λ' be rectangles with parallel sides such that $|\Lambda'| = nm\, |\Lambda|$, with n,m integers. By GKS and the translation invariance of periodic states we have

$$\exp(|\Lambda'|(\alpha^P_{\Lambda'}(\lambda+s) - \alpha^P_{\Lambda'}(\lambda))) = <\exp(s\phi(\chi_{\Lambda'})>^P_{\Lambda'} \geq$$

$$\geq <\exp(s\phi(\chi_\Lambda))>^{P^{nm}}_{\Lambda'} \geq <\exp(s\phi(\chi_\Lambda))>^{nm}_{\Lambda'}$$

where $\alpha^P_{\Lambda'}$ is the half-periodic pressure for which $\alpha^P_{\Lambda'} \to \alpha_\infty$ as $\Lambda' \to \infty$ [14]. Taking the limit as $\Lambda' \to \infty$ we establish the upper bound necessary for the theorem in the case $s \geq 0$. In order to establish equality in the limit for $s \geq 0$ let us remark that still by GKS we have

$$<\exp(s\phi(\chi_\Lambda))> \geq <\exp(s\phi(\chi_\Lambda))>_\Lambda = \exp|\Lambda|(\alpha_\Lambda(\lambda+s) - \alpha_\Lambda(\lambda))$$

Since the half-Dirichlet pressure α_Λ converges to α_∞ as $\Lambda \to \infty$ [13,14] the case $s \geq 0$ is complete.

On the other hand for $s \geq 0$ $<\sinh s\phi(\chi_\Lambda)>_{\Lambda'} \leq <\sinh s\phi(\chi_\Lambda)>$, and for any $\varepsilon > 0$ we can find d such that if Λ' is a rectangle whose boundary is at a distance greater than d from a fixed rectangle Λ we have

$$<\exp(s\phi(\chi_\Lambda))>_{\Lambda'} \geq <\exp(s\phi(\chi_\Lambda))>-\varepsilon$$

Therefore in these conditions

$$<\exp(-s\phi(\chi_\Lambda))>_{\Lambda'} \geq <\exp(-s\phi(\chi_\Lambda))>-\varepsilon$$

Let Λ_0 be a rectangle made of n^2 adjacent rectangles Λ_i equal to Λ, and Λ' a rectangle with sides parallel to those of Λ_0 and at a distance d from Λ_0. Then by FKG (not GKS!) we have

$$\exp|\Lambda'|(\alpha_{\Lambda'}(\lambda-s)-\alpha_{\Lambda'}(\lambda)) = <\exp(-s\phi(\chi_{\Lambda'}))>_{\Lambda'} \geq$$

$$\geq \Pi_i <\exp(-s\phi(\chi_{\Lambda_i}))>_{\Lambda'} \geq [<\exp(-s\phi(\chi_\Lambda))>-\varepsilon]^{n^2}$$

If we take $n^2 \to \infty$ so that $|\Lambda'|/n^2 \to |\Lambda|$ then we have immediately

$$|\Lambda|^{-1}\log < \exp(-s\phi(\chi_\Lambda))> \leq \alpha_\infty(\lambda-s) - \alpha_\infty(\lambda)$$

for $s \geq 0$ and Λ rectangle. The theorem is now complete.

Remark. We believe that a slight extension of our methods will give equality in theorem 21 also for $s < 0$.
Now we turn to the one point function.

Theorem 22. For the half-Dirichlet states $M(\lambda) = M^{(-)}(\lambda)$ for $\lambda > 0$. Obviously $M(0) = 0$ by symmetry.

Proof. Since for $f \geq 0$ $<\phi(f)>_\lambda$ can be obtained as a supremum of volume cut off states which are continuous in λ we have that $M(\lambda)$ must be lower semicontinuous and therefore by Theorem 8 must coincide with $M^{(-)}(\lambda)$ for $\lambda > 0$.

Remark. This result follows also from a comparison argument [22].

All result of Sections 4 and 5 transfer immediately to the half-Dirichlet states. But we have also

Theorem 23. For $\lambda \neq 0$ and any even polynomial Q the half-Dirichlet states $< >_\lambda$ are pure. Moreover $< >_\lambda = < >_\lambda^{(\pm)}$ for $\lambda < 0$ and $\lambda > 0$ respectively.

The proof follows easily from Theorems 11, 17 and 22.

In this way the problem of the ergodic decomposition of half-Dirichlet states is completely solved for $\lambda \neq 0$. Moreover, we believe that for $\lambda = 0$ the following holds

Conjecture 24. For any even polynomial Q and $\lambda = 0$

$$< > = \tfrac{1}{2}(< >^{(+)} + < >^{(-)})$$

Obviously this is trivially true if $< >^{(+)} = < >^{(-)}$ by Theorem 17.

References

[1] R. L. DOBRUSHIN, Gibbsian Random Fields for Lattice Systems with Pairwise Interactions, Funct. Anal. Applic. 2, 292 (1968).

[2] J. FELDMAN and K. OSTERWALDER, The Wightman Axioms and the Mass Gap for Weakly Coupled $(\phi^4)_3$ Quantum Field Theories, Preprint, Harvard University, February 1975.

[3] J. FELDMAN and K. OSTERWALDER, The Construction of $\lambda\phi_3^4$. Preprint, Harvard University, 1975.

[4] C. FORTUIN, P. KASTELEYN and J. GINIBRE, Correlation Inequalities on Some Partially Ordered Sets, Comm. Math. Phys. 22, 89 (1971).

[5] J. FRÖHLICH, Existence and Analyticity in the Bare Parameters of the $[\lambda(\vec{\phi}\cdot\vec{\phi})^2 - \sigma\phi_1^2 - \mu\phi_1]$ - Quantum Field Models, I. Princeton University Preprint.

[6] J. FRÖHLICH and B. SIMON, Preprint in preparation.

[7] J. GLIMM and A. JAFFE, The $\lambda(\phi^4)_2$ Quantum Field Theory without Cutoffs, III. The Physical Vacuum, Acta Math. 125, 203 (1970).

[8] J. GLIMM and A. JAFFE, Positivity of the ϕ_3^4 Hamiltonian, Fortschritte der Physik 21, 327 (1973).

[9] J. GLIMM, A. JAFFE and T. SPENCER, The Wightman Axioms and Particle Structure in the Quantum Field Model, Ann. Math. 100, 585 (1974).

[10] R. GRIFFITHS, C. HURST and S. SHERMAN, Concavity of Magnetization of an Ising Ferromagnet in a Positive External Field, J. Math. Phys. 11, 790 (1970).

[11] F. GUERRA, Uniqueness of the Vacuum Energy Density and Van Hove Phenomenon in the Infinite Volume Limit for Two-Dimensional Self-Coupled Bose Fields, Phys. Rev. Lett. 28, 1213 (1972).

[12] F. GUERRA, Exponential Bounds in Lattice Field Theory, Report at the Marseille Colloquium, June 1975.

[13] F. GUERRA, L. ROSEN and B. SIMON, The $P(\phi)_2$ Euclidean Quantum Field Theory as Classical Statistical Mechanics, Ann. of Math. 101, 111 (1975).

[14] F. GUERRA, L. ROSEN and B. SIMON, Boundary Conditions in the $P(\phi)_2$ Euclidean Quantum Field Theory, Preprint, 1975.

[15] O. LANFORD and D. RUELLE, Observables at Infinity and States with Short Range Correlations in Statistical Mechanics, Commun. Math. Phys. 13, 194 (1969).

[16] J. LEBOWITZ and E. PRESUTTI, Preprint in preparation.

[17] J. MAGNEN and R. SENEOR, The Infinite Volume Limit of the ϕ_3^4 Model, Preprint, Ecole Polytechnique, February 1975.

[18] E. NELSON, Probability Theory and Euclidean Field Theory, Contribution to [27].

[19] Y. M. PARK, Convergence of Lattice Approximations and Infinite Volume Limit in the $(\lambda\phi^4 - \sigma\phi^2 - \mu\phi)_3$ Field Theory, ZiF Preprint, Bielefeld 1975.

[20] D. RUELLE, Statistical Mechanics, Benjamin, New York (1969).

[21] E. SEILER and B. SIMON, Nelson's Symmetry and all That in the Yukawa_2 and $(\phi^4)_3$ Field Theories, Preprint, 1975.

[22] B. SIMON, The $P(\phi)_2$ Euclidean (Quantum) Field Theory, Princeton University Press, 1974.

[23] B. SIMON, Preprint in preparation.

[24] B. SIMON and R. GRIFFITHS, The $(\phi^4)_2$ Field Theory as a Classical Ising Model, Commun. Math. Phys. 33, 145 (1973).

[25] T. SPENCER, The Mass Gap for the $P(\phi)_2$ Quantum Field Model with a Strong External Field, Commun. Math. Phys. 39, 63 (1974).

[26] K. SYMANZIK, Euclidean Quantum Field Theory in Local Quantum Theory, R. Jost, Editor, Academic Press, New York (1969).

[27] G. VELO and A. Wightman, Constructive Quantum Field Theory, Lecture Notes in Physics, Vol. 25, Springer-Verlag, Berlin (1973).

Acta Physica Austriaca, Suppl. XVI, 147–166 (1976)

Critical Exponents and Renormalization in the ϕ^4 Scaling Limit

J. Glimm[+]
Rockefeller University
New York, New York 10021

A. Jaffe[++]
Harvard University
Cambridge, Mass. 02138

Abstract

For dimensions $d \leq 3$, the ϕ^4 scaling limit defines a nonrenormalizable field theory. The standard relations between critical exponents and renormalization are presented. Arguments supporting the existence of the scaling limit are based on correlation inequalities and the numerical values of Ising model exponents, $2\eta \underset{\neq}{<} \eta_E$ for d=2,3.

Contents

[+] Supported in part by the National Science Foundation under Grant 74 - 13252.

[++] Supported in part by the National Science Foundation under Grant 75 - 21212.

1. Formulation of the Problem

The Euclidean action density

$$:(\nabla\phi^2 + \lambda\phi^4 + \sigma\phi^2 + \mu\phi): \tag{1.1}$$

determines a quantum field theory for d=1,2,3 [7,21,22,4,20,26, 6,23]. The Wick order is defined with a bare mass 1. Dirichlet boundary conditions and (for d=3) additional mass counterterms are implied in (1.1). For large positive values of σ, the field has a unique state Ω and a positive mass

$$m = m(\lambda,\sigma,\mu) > 0, \tag{1.2}$$

[13,4,20]. For d=2 (and presumably for d=3), for $\mu=0$, and for large negative value of σ, the vacuum is degenerate [14]. We define the critical value of σ, $\sigma_c=\sigma_c(\lambda)$ as the supremum of the values of σ for which either the vacuum is degenerate, or for which (1.2) fails.

On the interval $(\sigma_c(\lambda),\infty)$, $m(\lambda,\sigma,0)$ is monotone increasing and Lipschitz continuous [16,9] in σ, for d=2 at least, and guided by theorems concerning lattice fields [1,24] (and arbitrary $d\geq 2$), we expect that $m\searrow 0$ as $\sigma\searrow\sigma_c$. This is the only mathematically rigorous statement which can be made about the critical point: for ϕ_2^4 fields (and Ising models), $\sigma_c < \infty$ while for lattice ϕ^4 fields (and Ising models) $m(\sigma)\searrow 0$ as $\sigma\searrow\sigma_c$. In addition to the corresponding continuum problem, as mentioned above, the uniqueness of the vacuum at $\sigma=\sigma_c$ and the conjecture

$$m(\lambda,\sigma_c,0) = 0$$

are open problems, both for Ising models ($d\geq 3$), and lattice and continuum fields ($d\geq 2$). The d=2 Ising model, I_2, is of course a special case, because the existence of a closed form solution makes the detailed critical structure accessible [2].

Because of the absence of a mathematical theory of critical behavior, the remainder of our discussion will be mainly on a heuristic level. Let $\langle\ldots\rangle = \int \ldots d\phi$ be the Euclidean vacuum expectation associated with the quantum field defined by (1.1). The two point Schwinger function

$$S^{(2)}(x,y) = \int \phi(x)\phi(y)d\phi$$

can be represented, according to the Lehmann spectral theorem, in the form

$$S^{(2)}(x,y) = \int C_a(x-y)d\rho_a, \qquad (x \neq y)$$

where C_a is the convolution inverse to $(-\Delta+a)$. Necessarily,

$$m^2 = \inf \operatorname{suppt} d\rho_a$$

and we make the further assumption (absence of bound states in $S^{(2)}$) that

$$S^{(2)}(x,y) = Z_3 C_{m^2}(x-y) + \int_{a \geq (3m)^2} C_a(x-y)d\rho_a \tag{1.3}$$

The evidence in favor of this assumption will be presented below. In particular, we take $\mu=0$, $\sigma>\sigma_c$, since otherwise bound states may be expected. Here the wave function renormalization constant Z_3 is defined by (1.3).

The problem of critical exponents is to understand the leading singularity of the long distance behavior of the field ϕ, as $\sigma \to \sigma_c$, in particular as expressed in such formulae as

$$m \sim \text{const} \left(\frac{\sigma-\sigma_c}{\sigma_c}\right)^{\nu} \tag{1.4}$$

$$\chi \equiv S^{(2)}(x-y)dx \sim \text{const} \left(\frac{\sigma-\sigma_c}{\sigma_c}\right)^{-\gamma} \tag{1.5}$$

$$z_3 \sim \text{const} \left(\frac{\sigma-\sigma_c}{\sigma_c}\right)^{\zeta_3} \tag{1.6}$$

$$S^{(2)}(x-y)\Big|_{\sigma=\sigma_c} \sim \text{const}\,|x-y|^{2-d-\eta} \quad \text{as } |x-y|\to\infty \tag{1.7}$$

2. The Scaling and Critical Point Limits

The ϕ^4 field theory, $d\leq 3$, has two intrinsic length scales. The longer length scale is the correlation length,

$$\xi = m^{-1}$$

and this length governs the long distance decay of the correlation functions

$$S^{(n)} = \int \phi(x_1)\ldots\phi(x_n)d\phi$$

In fact for n=2,

$$S^{(2)}(x-y) \sim Z_3 m^{d-2}(mr)^{(1-d)/2}e^{-mr}, \quad mr\to\infty$$

assuming (1.3). The short distance behavior of a ϕ^4 field is canonical (free) for $d\leq 3$, cf. [8,10]. The short length scale is the distance scale on which this canonical behavior becomes dominant. For a lattice field or Ising model, the short length scale could be the lattice spacing. The lattice spacing, of course, functions as an ultraviolet cutoff, but we will see below that the short distance scale is always an ultraviolet cutoff. In the case of a Lorentz covariant ϕ^4 continuum field theory, the short distance scale cuts off the nonrenormalizable infrared singularities, and substitutes a Lorentz covariant short distance canonical behavior, for d=2,3. It could happen (for example in the continuum limit of lattice theories) that the lattice spacing is very short relative to the above Gaussian short distance scale. This case, in which there are two short distance scales and one long distance scale, will not be discussed further.

In the case of two length scales, one can always be eliminated by a scale transformation, and only the dimensionless ratio

$$\varepsilon = \frac{\text{short distance scale}}{\text{long distance scale}} \tag{2.1}$$

has an intrinsic signifiance. The scale transformation U_s is a unitary operator connecting the Hilbert spaces of the fields (1.1) having the same value of dimensionless bare charge

$$g = \lambda\sigma^{(d-4)/2} \tag{2.2}$$

and differing values of σ. U_s is defined to multiply all lengths by the factor s, so that

$$U_s\phi(x)U_s^{-1} = s^{(2-d)/2}\phi(s^{-1}x) \equiv \psi_s(x) \tag{2.3}$$

defines a field with the new parameter values

$$\lambda_s = s^{d-4}\lambda, \qquad \sigma_s = s^{-2}\sigma$$

and Wick ordering mass s^{-1}. (A reWick ordering then gives the action of the field ψ_s the form (1.1).)

There are two distinguished choices for the parameter s, namely

$$s = 1 \qquad \text{or} \qquad s = \varepsilon$$

In the first choice, s=1, the short distance scale is fixed at one; this choice defines the <u>unscaled</u> theory. In the second choice, s=m, the long distance scale ξ = 1/mass, is fixed at one; this choice defines the <u>scaled</u> theory. We note ε is the mass of the unscaled theory and the lattice spacing or short distance cutoff of the scaled theory.

It follows that there are three distinguished distance intervals:

$$r \le \text{short distance scale} \tag{2.4a}$$

$$\text{short distance scale} \le r \le \text{long distance scale} \tag{2.4b}$$

$$\text{long distance scale} \le r \tag{2.4c}$$

We call these intervals (a) the canonical (or free, or Gaussian) interval, (b) the scaling (or critical) interval, and (c) the one particle interval. In the two extreme intervals, i.e. in the canonical and one particle intervals, the behavior of ϕ is well understood. In cases where these two intervals dominate, (high or low temperature) there is a convergent cluster expansion, and ϕ is well understood on all distance scales [13,14,15].

Near the critical point, ε becomes small, and the critical, or scaling interval (2.4b) dominates. As $\varepsilon \to 0$, we may construct three limiting theories.

i) The critical point is defined by $\varepsilon \to 0$, $\sigma \searrow \sigma_c$ in the unscaled theories. This limit is characterized by

$0 = m$

$0 \neq$ short distance scale.

ii) The scaling limit is defined by $\varepsilon \to 0$ in the scaled theories, and is characterized by

$1 = m$

$0 =$ short distance scale

In this limit, we use the renormalized field $\phi_{ren} = Z_3^{-1/2}\phi$ in place of ϕ.

iii) The scaled critical point is defined by a long distance scaling of i) or a short distance scaling, $m \to 0$, of ii), and is characterized by

$m = 0 =$ short distance scale

These two definitions of iii) are related by an interchange of order of limits.

Alternately, we may say that i) eliminates (2.4c), ii) eliminates (2.4a) and iii) eliminates both, in either order. To check this interchange in the order of limits, we analyze its influence on η, using Ising model exponents. In the limit iii),

$$S^{(2)} = r^{2-d-\eta} \tag{2.5}$$

and η as defined in (1.7), refers to the long distance scaling of i).

We now assert, as a scaling hypothesis, that the decay rate (2,5) holds for $S^{(2)}$ for the entire interval (2.4b). As a check on this assertion, we use it to compute (in the unscaled theory)

$$\chi = \int S^{(2)}(r)dr \sim \int_{|r|\le\varepsilon^{-1}} r^{2-d-\eta}dr \sim \varepsilon^{-2+\eta}$$

In (1.4-6), the unscaled theory is understood, and so

$$\varepsilon \sim \left(\frac{\sigma-\sigma_c}{\sigma_c}\right)^{\nu} ; \quad Z_3 \sim \varepsilon^{\zeta_3/\nu} ; \quad \chi \sim \varepsilon^{-\gamma/\nu}$$

Thus the consistency check follows from the identity

$$\gamma = (2-\eta)\nu \tag{2.6}$$

valid for Ising model exponents, d=1,2,3.

We next consider the scaled theory. At the short distance end of (2.4b),

$$S^{(2)}_{ren}(\varepsilon) = Z_3^{-1}S^{(2)}_{free}(\varepsilon) = Z_3^{-1}\varepsilon^{2-d} = \varepsilon^{2-d-\zeta_3/\nu} \tag{2.7}$$

neglecting logarithms in d=2 dimensions. Thus

$$\begin{aligned} S^{(2)}_{ren}(r) &= \varepsilon^{2-d-\zeta_3/\nu}(r/\varepsilon)^{2-d-\eta} \\ &= \varepsilon^{\eta-\zeta_3/\nu} r^{2-d-\eta} \end{aligned} \tag{2.8}$$

on (2.4b) by the scaling hypothesis. We note that the (rigorous) inequalities

$$2\nu - \zeta_3 \le \gamma \lesssim 2\nu - \eta\nu$$

[5,9] imply that

$$\eta - \zeta_3/\nu \le 0 \tag{2.9}$$

so that (2.8) does not vanish as $\varepsilon\to 0$.

We further require equality in (2.9), as is known for I_2, and as follows from the existence of the limit ii), in the special case (2.8). Then in the scaling limit ii), we have

$$S^{(2)}_{ren}(r) \cong r^{2-d-\eta}, \quad r << 1 \tag{2.10}$$

and scaling from short distances gives $S^{(2)}_{ren}(r) = r^{2-d-\eta}$ for all r. Thus the two definitions of the limit iii) agree, at least in their two point function. In particular the two definitions of η agree. In summary we can say that interchange of limits in iii) follows from a scale relation, such as (2.5), valid over the entire scaling interval (2.4b), and that this hypothesis can be checked against known values of the exponents.

A comprehensive exposition of the scaling behavior and the theory of the renormalization group is given in [28].

3. Renormalization of the $\phi^2(x)$ Field

The field $\phi^2(x)$ requires both an additive and a multiplicative renormalization. The additive renormalization is Wick ordering in the physical vacuum, defined by

$$:\phi^2(x): \;=\; \phi^2(x) \;-\; \langle\phi^2(x)\rangle \tag{3.1}$$

(These Wick dots do not coincide with those of (1.1), but in the term $\sigma\int:\phi^2:$ the difference is a constant, and of no consequence). The multiplicative renormalization

$$\begin{aligned}(:\phi^2(x):)_{\text{ren}} &= Z_E^{-1/2}:\phi^2(x):\\ &\neq \;:\;(\phi_{\text{ren}})^2(x)\;:\end{aligned} \tag{3.2}$$

is not obtained by renormalizing each factor ϕ in $\phi^2(x)$. Rather we define the exponents

$$\langle:\phi^2(x)::\phi^2(y):\rangle\Big|_{\sigma=\sigma_c} \sim \text{const}\,|x-y|^{4-2d-\eta_E} \quad \text{as} \quad |x-y|\to\infty \tag{3.3}$$

$$C_H = \int\langle:\phi^2(x)::\phi^2(y):\rangle d(x-y) \sim \left(\frac{\sigma-\sigma_c}{\sigma_c}\right)^{-\alpha} = \varepsilon^{-\alpha/\nu} \tag{3.4}$$

Here C_H is the specific heat, and if $Z(\sigma,\mu)$ is the partition function for (1.1), then

$$C_H = \frac{\partial^2}{\partial\sigma^2}\,\frac{\ln Z}{V}$$

$$\chi = \frac{\partial^2}{\partial\mu^2}\,\frac{\ln Z}{V}$$

As in §2, we make a scaling hypothesis, that (3.3) is valid over the scaling interval (2.4b), and then the identity in the unscaled theory

$$\varepsilon^{-\alpha/\nu} = \int_{1\leq|r|\leq\varepsilon^{-1}} r^{4-2d-\eta_E}dr + \int_{|r|\leq 1} r^{4-2d}dr$$

$$= \varepsilon^{-(4-d-\eta_E)}$$

implies

$$\alpha/\nu = 4-d-\eta_E = \begin{matrix} 0, & d=2 & \text{Ising} \\ 2 & d=3 & \text{Ising} \end{matrix} \tag{3.5}$$

This identity is analogous to the identity (2.6) for the ϕ-two point function. As in §2, a matching of the canonical form of the ϕ^2 two point function on (2.4a) with its scaling form on (2.4b) implies $\zeta_E/\nu = \eta_E$. In fact we assume in the unscaled theory that

$$\langle:\phi^2(x)::\phi^2(y):\rangle \sim c_a r^{4-2d} \qquad \text{on (2.4a)}$$

$$\sim c_b(\varepsilon) r^{4-2d-\eta_E} \qquad \text{on (2.5b)}$$

Because $\sigma=\sigma_c$ is a regular point for the canonical part of ϕ, we may take $c_a=c_a(\varepsilon)\big|_{\varepsilon=0}$ to be independent of ε. Equality of the two asymptotic forms at r=1 implies $c_b=c_a$ is also regular in ε. It follows that in the scaled theory

$$\langle:\phi^2(x)::\phi^2(y):\rangle \sim c_b \varepsilon^{4-2d}(r/\varepsilon)^{4-2d-\eta_E} \tag{3.6a}$$

and

$$\langle:\phi^2:_{ren}:\phi^2(y):_{ren}\rangle \sim c_b \varepsilon^{\eta_E-\zeta_E/\nu} r^{4-2d-\eta_E} \tag{3.6b}$$

on (2.4b). We now define $Z_E = \varepsilon^{\zeta_E/\nu}$ and

$$\zeta_E/\nu = \eta_E \tag{3.7}$$

The fact that

$$\eta_E = \begin{cases} 2, & d=2 \\ .8, & d=3 \end{cases} > 2\eta = \begin{cases} 1/2, & d=2 \\ .082, & d=3 \end{cases} \tag{3.8}$$

implies the following important result:

$$:(\phi_{ren})(x)^2: = 0 \tag{3.9}$$

in the scale limit. Here we obtain the values for η_E from (3.5).

Theorem 3.1. For the scaled theories,

$$||\int :(\phi_{ren})(x)^2:dx||^2_{L_2} \sim \varepsilon^{\eta_E - 2\eta}$$

Proof. We use (3.6)-(3.7) as follows:

$$||\int_\Lambda :(\phi_{ren})(x)^2:dx||^2_{L_2} = \varepsilon^{\eta_E - 2\eta}||\int_\Lambda :\phi(x)^2:_{ren}dx||^2_{L_2}$$

$$= O(1)\varepsilon^{\eta_E - 2\eta} \int_{\varepsilon\le|r|\le 1} r^{4-2d-\eta_E}dr$$

$$= O(1)\varepsilon^{\eta_E - 2\eta}$$

In more intuitive language, $:(\phi_{ren})^2:$ and $:\phi^2:_{ren}$ differ by an infinite multiple. Since the constant Z_E is defined so that the larger, $:\phi^2:_{ren}$, is finite, the smaller, $:(\phi_{ren})^2:$, must vanish.

4. Existence of the Scaling Limit

Existence of the scaling limit, in the weakest sense, means bounds uniform in ε on the renormalized Schwinger functions

$$S^{(n)}_{ren} = Z_3^{-n/2} S^{(n)}$$

so that by compactness, a convergent subsequence may be selected. By the explicit introduction of an invariant mean, the limit may be taken to be covariant under the translation subgroup. The Osterwalder-Schrader reconstruction theorem [21,22] then guarantees the existence of a scaling limit field theory which is at least translation covariant.

The Lebowitz correlation inequalities bound $S^{(n)}_{ren}$ by a sum of products of two point functions [11]. Thus existence follows from a uniform bound on $S^{(2)}_{ren}$. The required bounds on $S^{(2)}_{ren}$ are implied [12] by a conjectured correlation inequality

$$\Gamma^{(6)}(xxx\ yyy) \leq 0 \tag{4.1}$$

Here $\Gamma^{(6)}(x_1,\ldots,x_6)$ is the six point vertex, or direct correlation function. In graphical language, $\Gamma^{(6)}$ is one particle irreducible, and in (4.1) we may take $\Gamma^{(6)}$ in either its unamputated or its amputated form. Let $G^{(n)}$ denote the n-point truncated (Ursell) function. Choosing the unamputated form for the Γ's, and defining

$$\Gamma^{(2)} = -(G^{(2)})^{-1}$$

(convolution inverse), we have the explicit formulae

$$\Gamma^{(4)} = G^{(4)}$$

$$\Gamma^{(6)} = G^{(6)} + \frac{1}{2} \sum_{\text{Permutations}} (3!)^{-2} G^{(4)}(x_{i_1},\ldots x_{i_3},z) \times$$

$$\Gamma^{(2)}(z,z')G^{(4)}(a',x_{i_4},\dots x_{i_6})$$

The $G^{(n)}$ are connected parts of the $S^{(n)}$.

The identity $\phi^2 \equiv$ const. in the scaling limit, allows an explicit calculation of $\Gamma^{(6)}(\mathrm{xxx\ yyy})$. We first consider the more elementary calculation

$$G^{(4)}(xxyy) = S^{(4)}(xxyy) - S^{(2)}(xx)S^{(2)}(yy) - 2S^{(2)}(xy)^2$$

$$= \int(\phi^2(x) - \langle\phi^2(x)\rangle)(\phi^2(y) - \langle\phi^2(y)\rangle)d\phi - 2S^{(2)}(xy)^2$$

$$= \int :\phi^2(x)::\phi^2(y): d\phi - 2S^{(2)}(xy)^2 = -2S^{(2)}(xy)^2$$

For $\Gamma^{(6)}$ we have

$$\Gamma^{(6)}(xxxyyy) = -24\ G^{(2)}(xy)^3 \tag{4.2}$$

(4.2) holds only for $\varepsilon=0$, and thus does not establish (4.1) as $\varepsilon\to 0$, but it certainly makes the inequality (4.1) highly plausible, at least for small ε. Other tests of (4.1) include the one-dimensional Ising model, in which

$$\Gamma^6(x_1,\dots,x_6) = -24\ G^{(2)}(x_1x_6)G^{(2)}(x_2x_5)G^{(2)}(x_3x_4)$$

for

$$x_1 \leq x_2 \leq \dots \leq x_6$$

[25], and numerical studies of the d=1 ϕ^4 field theory -- i.e., the anharmonic oscillator [19].

We remark that (4.1) has one other consequence: an absence of bound states in the two point function [12]. In particular if (4.1) holds, then $S^{(2)}$ has the form (1.3), with $0<Z_3\leq 1$. Moreover in case (4.1) holds, there are no CDD zeros, which means that $\Gamma^{(2)} = -(G^{(2)})^{-1}$ has a 3m decay rate [12]. In particular the scaling limit has neither three particle bound states nor CDD zeros. Even bound states are also excluded by a general argument [3,27], which also applies for $\sigma>\sigma_c$.

The absence of bound states may be special to $\sigma>\sigma_c$, $\mu=0$. In fact for the two dimensional Ising model with $T<T_c$, it is suggested [15] that as $\mu \searrow 0$, the number of bound states becomes infinite, and fills in to form a continuum threshold. Here we appeal to the explicit solution of the Ising model for $T<T_c$, $\mu=0$. If the Ising and ϕ^4 theories have identical scaling limits, then a similar phenomena should occur in ϕ^4 field theory for $\sigma \lesssim \sigma_c, \mu \searrow 0$.

For d=1, existence of the scaling limit was established [17] on the basis of asymptotic estimates on the eigenvalues and eigenvectors of the anharmonic oscillator. This scale limit coincides with the one dimensional Ising model, proving universality in this case.

5. The Josephson Inequality

The Josephson inequality states

$$2-\alpha \leq d\nu \qquad (5.1)$$

[18]. For I_2, (5.1) is an equality, and for I_3, (5.1) is correct as an inequality, and nearly (or possibly) correct as an equality. For the Gaussian (free field) case, the exponents are

$$\nu = 1/2, \quad \alpha = 2-d/2, \quad 0 = \eta = \eta_E = \zeta_3 = \zeta_E, \quad \gamma = 1$$

and (5.1) is again an equality.

Following a similar calculation in [9] based on Lebowitz's inequality, we have

$$\varepsilon^{-\alpha/\nu} \sim C_H \leq \int \langle \phi(x)\phi(y)\rangle^2 dx$$

$$\leq O(1) \int_{|r|\leq \varepsilon^{-1}} r^{4-2d-2\eta} dr$$

$$\leq O(\varepsilon^{-(r-d-2\eta)}) \qquad \text{for } d = 2,3,4.$$

Thus

$$\alpha \leq (4-d-2\eta)\nu \qquad d = 2,3,4. \qquad (5.2)$$

Here we have used the scaling hypothesis for the two point function. Without this hypothesis, the same reasoning implies $\alpha \leq (4-d)\nu$ as a rigorous inequality, d=2,3.

Combined with (5.1), (5.2) reproduces the inequality $\nu \geq (2-\eta)^{-1} \geq \frac{1}{2}$ Furthermore of α/ν is strictly less than its Gaussian value, 4-d, then ν is strictly greater than its canonical value, 1/2. From Lebowitz's inequality, we have the (rigorous) inequality

$$\eta_E \geq 2\eta \geq 0 = \eta_{E\ \text{Gaussian}} = \eta_{\text{Gaussian}} \tag{5.3}$$

for d=1,2,3.

We complete this section with a formal derivation of (5.1), based on field theory ideas. The Hamiltonian and particle structure occur as hypotheses, and use is made of Lebowitz inequality. From (1.4), we have

$$\frac{dm}{d\sigma} \sim \left(\frac{\sigma-\sigma_c}{\sigma_c}\right)^{\nu-1} = \varepsilon^{1-1/\nu}$$

Let $<1|$ be the one particle zero momentum state. Then

$$m = <1|H|1>$$

and

$$\frac{dm}{d\sigma} = <1|\frac{dH}{d\sigma}|1>$$

Equating these two expressions gives

$$\varepsilon^{1-1/\nu} \sim <1|\int :\phi^2(x):d\vec{x}|1>$$

and because $m \int_0^{1/m} e^{-mt}dt = 1-e^{-1}$ and $m = \varepsilon$,

$$\varepsilon^{-1/\nu} \sim <1|\int_{|x_o|\leq m^{-1}} :\phi^2(x):dx|1> \tag{5.4}$$

The right side of (5.4) will be bounded by a Schwarz inequality. To justify this step, we use two approximations

$$<1| \to \text{const} \int_{|x|\leq m^{-1}} \phi(x)d\vec{x}\Omega \tag{5.5}$$

$$\text{const} = \Big(\int_{\substack{|\vec{x}|\leq m^{-1} \\ |\vec{y}|\leq m^{-1}}} S^{(2)}(x-y)d\vec{x}\ d\vec{y}\Big)^{-1/2}$$

and

$$\int_{|x_o|\leq m^{-1}} :^2(x):dx \to \int_{|x|\leq m^{-1}} :\phi^2(x):dx \tag{5.6}$$

With these approximations, $\langle 1|\varepsilon L_4$, with an L_4 norm bounded independently of ε. In fact $||\langle 1|||_{L_4}^4$ is a four point function divided by a product of two two point functions. The four point function can be expanded as

$$S^{(4)} = G^{(4)} - \sum G^{(2)} G^{(2)}$$

and by Lebowitz inequality

$$0 \leq -G^{(4)} \leq G^{(2)} G^{(2)}$$

Here the two point functions $G^{(2)}$ are bounded by the two point functions occurring in the denominator, and so $\langle 1|\varepsilon L_4$ as asserted.

Applying the Schwarz inequality to (5.4) yields

$$\varepsilon^{-1/\nu} \leq O\Big(\int_{|x|,|y| \leq m} \langle :\phi^2(x)::\phi^2(y):\rangle dx\, dy\Big)^{1/2}$$

$$\sim (\varepsilon^{-d-\alpha/\nu})^{1/2}$$

Thus

$$1/\nu \leq \tfrac{1}{2}(d + \alpha/\nu)$$

and (5.1) follows.

Because of the wave function renormalization in the scaling limit, we include a discussion of the substitution (5.5). Let $\langle 1_{appx}|$ denote the right side of (5.5). From the scaling hypothesis, we find that

$$\text{const} = O(\varepsilon^{(d+2-\eta)/2}) = Z_3^{-1/2} O(\varepsilon^{d/2})$$

in (5.5). The factor $Z_3^{-1/2}$ replaces ϕ by ϕ_{ren}, and ensures that the one particle contribution to ϕ_{ren} has unit strength. The factor is proportional to

$$\Big|\Big| \int_{|\vec{x}| \leq m^{-1}} \phi_{free}(x)\, d\vec{x}\,\Omega_{free} \Big|\Big|^{-1}$$

and guarantees that the one particle portion of $\langle 1_{appx}|$ has strength $O(1)$, uniformly as $\varepsilon \to 0$.

References

1. G. BAKER, Self-interacting boson quantum field theory and the thermodynamic limit in d dimensions, J. Math. Phys. 16, 1324 - 1346 (1975).
2. E. BAROUCH, B. MCCOY, C. TRAY and T. T. WU, The spin-spin correlation functions for the two dimensional Ising-model Exact theory in the scaling limit, to appear.
3. J. FELDMAN, On the absence of bound states in the ϕ^4 quantum field model without symmetry breaking, Cand. J. Phys. 52, 1583 - 1587 (1974).
4. J. FELDMAN and K. OSTERWALDER, The Wightman axioms and the mass gap for weakly coupled $(\phi)^4_3$ quantum field theories, to appear.
5. M. FISHER, Rigorous inequalities for critical point correlation exponents, Phys. Rev. 180, 594 - 600 (1969).
6. J. FRÖHLICH, Existence and analyticity in the bare parameters of the $|\lambda(\vec{\phi}\,\vec{\phi})^2 - \sigma\phi_1^2 - \mu\phi_1|$ quantum field models, I. Manuscript.
7. J. GLIMM and A. JAFFE, The $(\lambda\phi^4)_2$ quantum field theory without cutoffs III. The physical vacuum, Acta Math. 125, 203 - 261 (1970).
8. ———, The $(\lambda\phi^4)_2$ quantum field theory without cutoffs IV. Perturbation of the Hamiltonian, J. Math. Phys. 13, 1558 - 1584 (1972).
9. ———, ϕ^4_2 quantum field theory in the single phase region: Differentiability of the mass and bounds on critical exponents, Phys. Rev. D10, 536 - 539 (1974).
10. ———, Two and three body equations in quantum field models, Commun. Math. Phys. 44, 293 - 320 (1974).
11. ———, A remark on the existence of ϕ^4_4. Phys. Rev. Lett. 33, 440 - 442 (1974).
12. ———, Three particle structure of ϕ^4 interactions and the scaling limit, Phys. Rev. D11, 2816 - 2827 (1975).
13. J. GLIMM, A. JAFFE, and T. SPENCER, The particle structure of the weakly coupled $P(\phi)_2$ models and other applications of high temperature expansions. In: Constructive quantum field theory, G. Velo and A. Wightman (eds.) Springer Verlag, Berlin, 1973.

14. J. GLIMM, A. JAFFE, and T. SPENCER, Existence of phase transitions for ϕ_2^4 quantum fields, Commun. Math. Phys. To appear.
15. ———————, A cluster expansion for the ϕ_2^4 quantum field theory in the two phase region. In preparation.
16. F. GUERRA, L. ROSEN, and B. SIMON, In: Constructive quantum field theory, G. Velo and A. Wightman (eds.) Springer Verlag, Berlin, 1973.
17. D. ISAACSON, The critical behavior of the autoharmonic oscillator, NYU Thesis.
18. B. JOSEPHSON, Inequality for the specific heat, I Derivation, II Applications, Proc. Phil. Soc. 92, 269 - 284 (1967).
19. D. MARCHESIN, Work in progress.
20. J. MAGNEN and R. SENEOR, The infinite volume limit of the ϕ_3^4 model, Ann. Inst. H. Poincaré, to appear.
21. K. OSTERWALDER and R. SCHRADER, Axioms for Euclidean Green's functions, Commun. Math. Phys. 31, 83 - 112 (1974).
22. ———————, Axioms for Euclidean Green's functions II, Commun. Math. Phys. 42, 281 - 305 (1975).
23. Y. PARK, Uniform bounds of the pressure of the $\lambda\phi_3^4$ field model. Preprint.
24. J. ROSEN, Mass renormalization for $\lambda\phi_2^4$ Euclidean lattice field theory.
25. J. ROSEN, Private communication.
26. E. SEILER and B. SIMON, Nelson's symmetry and all that in the Yukawa_2 and ϕ_3^4 field theories. Preprint.
27. T. SPENCER, The absence of even bound states in ϕ_2^4. Commun. Math. Phys. 39, 77 - 79 (1974).
28. E. BREZIN, J. C. LEGUILLORE and J. ZINN-JUSTIN, Field theoretical approach to critical phenomena. In: Phase transitions and critical phenomena, Vol. VI., Ed. by Domb and Green, Academic Press, New York, to appear.

Acta Physica Austriaca, Suppl. XVI, 167–175 (1976)

An Asymptotic Perturbation Expansion for Multiphase ϕ_2^4.

Arthur Jaffe**
Harvard University
Cambridge, Mass. 02138

James Glimm*
Rockefeller University
New York, N.Y. 10021

Thomas Spencer*,**
Rockefeller University
New York, N.Y. 10021

* Research supported in part by the National Science Foundation under Grant MPS74-13252.

** Research supported in part by the National Science Foundation under Grant MPS75-21212.

We consider the d=2 quantum field, with interaction density $P(x) = \lambda x^4 - \frac{3}{4} x^2$. For $\lambda \ll 1$, this model is known to have a phase transition [1], i.e. there exist at least two different field theories for a given λ. The new result we announce here is a detailed investigation of the two theories in which ϕ is concentrated near $\pm(8\lambda)^{-1/2}$, namely the global minima of $P(x)$. (Presumably these are the only pure states for the model.) We give a convergent expansion about the mean field approximation, which is asymptotic in powers of $\lambda^{1/2}$. We explain these ideas in more detail.

Let $d\mu_o$ denote the Gaussian probability measure on $\mathcal{S}'_{real}(R^2)$ with mean zero and covariance $(-\Delta+1)^{-1}$. The ground state measure for the $P(\phi)_2$ model is defined as

$$d\mu = \lim_{\Lambda \nearrow R^2} d\mu_\Lambda \tag{1}$$

where

$$d\mu_\Lambda = Z(\Lambda)^{-1} e^{-\int_\Lambda :P(\phi(x)):\, dx}\, d\mu_o. \tag{2}$$

Here the Wick ordering denoted by : : is defined with respect to the Gaussian measure $d\mu_o$ and $Z(\Lambda)$ is a normalizing constant chosen so that $\int d\mu_\Lambda = 1$. The limit in (1) exists in the sense of limit of moments or the limit of characteristic functions

$$S_\Lambda\{f\} = \int e^{i\phi(f)} d\mu_\Lambda \to S\{f\} = \int e^{i\phi(f)} d\mu.$$

In [1], we proved that for $P(x) = \lambda x^4 - \frac{3}{4} x^2$, with λ sufficiently small, the measure $d\mu$ of (1) is not ergodic under the action of the translation group $\phi(.) \to \phi(.-a)$ on R^2. In other words, $d\mu$ has a nontrivial decomposition into translation invariant, ergodic components: the pure phases. This is in contrast to the model with a positive quadratic term in the energy density: $\lambda x^4 + \frac{1}{2} x^2$, $\lambda << 1$. In this case d given by (1) is ergodic [3].

The physical picture suggesting this behavior is quite simple. The polynomial $P_1(x) = P(x) + \frac{1}{2} x^2$ is the classical approximation to the energy density. (The term $\frac{1}{2} x^2$ is the contribution from the covariance of the Gaussian measure $d\mu_o$.) With our choice of parameters ($\lambda << 1$, Wick ordering in $d\mu_o$) the quantum corrections to the energy density $P_1(x)$ vanish as $\lambda \to 0$. Thus the curve $P_1(x)$ pictured below gives a good qualitative idea of the energy density, and can be used as the lowest order term in an expansion about $\lambda = 0$.

$P_1(x) = \lambda x^4 - \frac{1}{4} x^2$

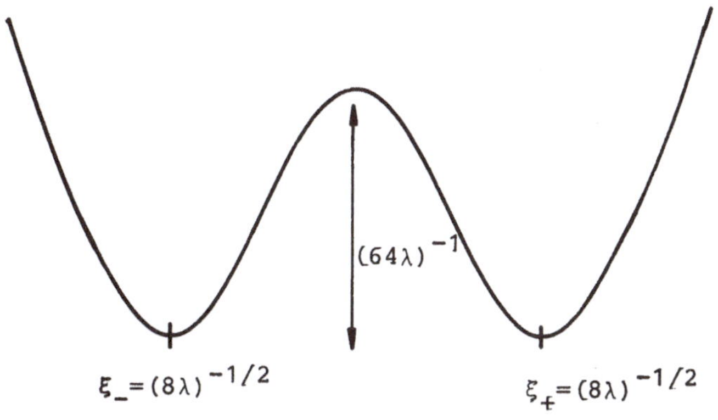

Furthermore, since the global minima $\xi_\pm$ of $P_1(x)$ become deeper and further separated as $\lambda \to 0$, we expect the measure $d\mu_\Lambda$ to be concentrated with greater and greater probability near $\phi = \xi_\pm = \pm(8\lambda)^{-1/2}$, as $\lambda \to 0$.

Translating to the minimum ξ_+, we find

$$Q(x) = P_1(x+\xi_+) = P_c(x) + (2\lambda)^{1/2} x^3 + \lambda x^4, \tag{3}$$

where the classical energy density and mass are defined by

$$\begin{aligned} P_c(x) &= \frac{1}{2} x^2 - (64\lambda)^{-1} \\ &= \frac{1}{2} m_c^2 x^2 + E_c . \end{aligned} \tag{4}$$

The quadratic polynomial $P_c(x)$ by itself would yield a translated Gaussian measure with mean ξ_+ and covariance $(-\Delta+1)^{-1}$. The coefficients of $Q(x) - P_c(x)$ are $O(\lambda^{1/2})$ and $O(\lambda)$, and thus we expect small deviation from this Gaussian if $\lambda << 1$ and $\phi \simeq \xi_+$. Likewise, expanding about ξ_- leads us to expect a Gaussian with mean ξ_-, plus small corrections, in case $\phi \simeq \xi_-$. In summary, the classical (or mean field) approximation to $P_1(x)$ suggests that $d\mu_\Lambda$, for $\lambda << 1$, is close to a direct sum of two Gaussian measures concentrated near $\xi_\pm$. Furthermore, we have established in [1] that the $\Lambda \nearrow R^2$ limit is not ergodic for $\lambda << 1$. Thus the mean field picture suggests that d splits into two components, $d\mu_\pm$, with $d\mu_\pm$ approximately given by the Gaussian $d\mu_{o,\pm}$, namely the measure $d\mu_o$ translated to $\xi_\pm$ respectively.

In our present work we impose boundary conditions on $d\mu_\Lambda$ which destroy the $\phi \to -\phi$ symmetry and favor one minimum of P_1 (say ξ_+). We then find that the limit (1) exists and that the resulting $d\mu_+$ is ergodic. Furthermore the moments of $d\mu_+$ have an asymptotic expansion in powers of $\lambda^{1/2}$, the leading term being the moments of $d\mu_{o,+}$.

In more detail, we define the boundary conditions as follows: Let

$$R(\Lambda) = \int_\Lambda : P_1(\phi(x)) : - \frac{1}{2} \int_\Lambda : (\phi-\xi_+)^2 : dx \tag{5}$$

$$= \int_\Lambda : \lambda\phi(x)^4 - \frac{1}{4}\phi(x)^2 - \frac{1}{2}(\phi-\xi_+)^2 : dx,$$

with Wick ordering defined by $d\mu_o$. Let

$$d\mu_\Lambda = \frac{1}{Z(\Lambda)} e^{-R(\Lambda)} d\mu_{o,+} . \qquad (6)$$

Thus the $(\phi-\xi_+)^2$ term in (5) exactly compensates for the mass and mean of $d\mu_{o,+}$ inside Λ, but leaves the translated measure (boundary condition) on the complement of Λ.

Theorem 1.

For $0 < \lambda$ sufficiently small, and $d\mu_\Lambda$ given by (6),

$$\lim_{\Lambda \nearrow R^2} d\mu_\Lambda = d\mu_+$$

exists in the sense of convergence of moments. The limiting $d\mu$ defines a theory satisfying the Osterwalder-Schrader and the Wightman axioms. The measure $d\mu_+$ is ergodic with respect to translations and has an exponential mixing rate (exponential clustering). Thus the Wightman theory has a unique vacuum and a positive mass.

In order to make precise the sense in which the moments of $d\mu_+$,

$$S_n(x_1,\ldots,x_n) = \int \phi(x_1)\ldots\phi(x_n)d\mu_+,$$

approximate the moments of $d\mu_{o,+}$, we consider the expansion in powers of $\lambda^{1/2}$ of the $\lambda\phi^4+(2\lambda)^{1/2}\phi^3$ perturbation of $d\mu_o$. Let

$$\langle\phi(x_1)\ldots\phi(x_n)\rangle_r \equiv \lim_{\Lambda\nearrow R^2}\left(\frac{\int\phi(x_1)\ldots\phi(x_n)\, e^{-\int_\Lambda i\lambda\phi^4+(2\lambda)^{1/2}\phi^3 : dx} d\mu_o}{\int e^{-\int_\Lambda : \lambda\phi^4 + (2\lambda)^{1/2}\phi^3 : dx} d\mu_o}\right)_r \qquad (7)$$

where $(\quad)_r$ denotes the sum of terms of order $\lambda^{j/2}$, $j \leq r$, in the asymptotic expansion of the finite volume expectation.

(The r=0 term is just the free field.)

Theorem 2.

Let $n,r \in Z_+$ be given and let λ be sufficiently small. Then

$$\int (\phi(x_1)-\xi_+)\ldots(\phi(x_n) - \xi_+)d\mu_+ = \langle\phi(x_1)\ldots\phi(x_n)\rangle_r + O(\lambda^{(r+1)/2}). \tag{8}$$

Note, in particular, that the translated fields have moments which are continuous as $\lambda \to 0$. For n=1, explicit computation based on (8) shows that

$$\int \phi(x) d\mu_+ = \xi_+ + \alpha\lambda^{3/2} + O(\lambda^2), \tag{9}$$

where $\xi_+ = (8\lambda)^{-1/2}$ and where the constant

$$\alpha = 144\sqrt{2}\int C(x)^3 dx - 864\sqrt{2}\int C(x)C(y)C(x-y)^2\, dxdy$$

is the first quantum correction to ξ_+. Here $C(x-y)$ is the kernel of $(-\Delta+1)^{-1}$. In terms of Feynman diagrams, α is given by the diagrams

We also remark that (by symmetry) there is a measure $d\mu_-$ related to $d\mu_+$ by the automorphism $\phi \to -\phi$, and which also gives a pure phase, massive theory.

The estimates which yield convergence can be extended to non zero external field ($|\mu| < \lambda^2$) and to complex λ ($\mathrm{Im}\lambda < \varepsilon\, \mathrm{Re}\lambda$). For $\mu \neq 0$, only one state has been constructed, and this state converges to $d\mu_+$ as $\mu \searrow 0$ or to $d\mu_-$ as $\mu \nearrow 0$. The question of whether a metastable state exists is open. In other words does a second state exists for $\mu > 0$, obtained for example, by analytic continuation through $\mu = 0$ of the state for $\mu < 0$? Such a metastable state would decay under external perturbation, and its existence appears unlikely.

There are several basic ideas which go into the proof of these results. First we require refined estimates for the vacuum energy. We establish these estimates by geometric consideration of the polynomial $P_1(\xi)$ and an analysis of $Z(\Lambda)$ as a sum of terms according to whether contributions arise from ϕ taking values (a) near the minima of $P_1(.)$, (b) near the barrier at $P_1(0)$ or (c) for large values of $|\phi|$. Having established these estimates uniformly for $\lambda << 1$ [2], we can then make three expansions.

The first expansion is perturbation theory up to order r, about the minimum ξ_+ of P_1. This expansion yields the finite volume analog of the term $< >_r$ in (8), and gives an explicit form for the remainder (in a finite volume Λ). The perturbation expansion is standard, and provides the explicit form of deviation from the Gaussian (up to order r in the coupling $\lambda^{1/2}$). It requires no further explanation.

The second expansion is an "expansion in phase boundaries", which shows that the transition from $\phi \simeq \xi_+$ at one point to the value $\phi \simeq \xi_-$ nearby , has very small probability. In other words, a "local phase transition" is unlikely. In order to make this concept quantitive, we parameterize the states of the ϕ field model by mapping ϕ onto an Ising spin variable σ. Let $\{\Delta\}$ denote a cover of R^2 by unit lattice squares, and let

$$\sigma(x) \equiv \operatorname{sgn} \int_\Delta \phi(y)dy, \quad x \in \Delta \tag{10}$$

Thus $\sigma(x)$ specifies whether the average value of ϕ, near x, is positive or negative. Since the field, with high probability takes values near $\pm(8\lambda)^{-1} = \xi_\pm$, the parameterization by σ is sufficient for $\lambda << 1$. We then carry out a version of the Peierls' argument. Let us divide path space into $2^{|\Lambda|}$ subsets on which $\sigma(x)$ takes a particular value (± 1 on each $\Delta \subset \Lambda$). Each subset is specified by its contour Γ, or the set of boundary lines between squares with different values of $\sigma(x)$. We show that

$$\Pr(\Gamma) = \frac{Z(\Lambda,\Gamma)}{Z(\Lambda)} = \frac{Z(\Lambda,\Gamma)}{\sum_{\Gamma'} Z(\Lambda,\Gamma')} \leq e^{-\lambda^{-1/2}|\Gamma|},$$

for $\lambda << 1$. Here $Z(\Lambda,\Gamma)$ denotes the restriction of the integral defining $Z(\Lambda)$ to the subset of path space specified by the contour Γ. Using such estimates, we show that the small effects with a contour present ($\Gamma \neq \emptyset$) are too small to contribute to an asymptotic expansion about $\lambda = 0$.

Finally, we make a cluster expansion, as in [3], to pass to the $\Lambda = R^2$ limit. The cluster expansion reduces estimates uniform in Λ to finite volume estimates, which then yield the proof of the theorems.

We remark that these techniques work in principle for $d > 2$, and for $d = 2$, a general $P(\phi)$ model in the mean field region maps onto a higher spin Ising variable.

References

1. J. GLIMM, A. JAFFE and T. SPENCER, Phase Transitions for ϕ_2^4 Quantum Fields, Commun. Math. Phys. 45, 203-216 (1975).

   ____________________, Existence of Phase Transitions for ϕ_2^4 Quantum Fields, Proc. of the Colloq. on Mathematical Methods of Quantum Field Theory, Marseille, June 1975.

2. ____________________, A Convergent Expansion about Mean Field Theory: I. The Expansion. II. Convergence of the Expansion, to appear.

3. ____________________ , The Wightman Axioms and Particle Structure in the $P(\phi)_2$ Quantum Field Model, Ann. Math. 100, 585-632 (1974).

   ____________________, The Particle Structure of the Weakly Coupled $P(\phi)_2$ Model and other Applications of High Temperature Expansions. Part II: The Cluster Expansion. In "Constructive Quantum Field Theory", G. VELO and A. WIGHTMAN (eds.) Springer Lecture Notes in Physics, Volume 25, Heidelberg 1973.

Acta Physica Austriaca, Suppl. XVI, 177-184 (1976)

Renormalization Problem in Massless $(\phi^4)_{4+\varepsilon}$ Theory

K. Symanzik

DESY, Hamburg

This is a summary of recent work /1/, an abbreviated version of which was presented at this year's GIFT seminar /2/. Some preliminary results were communicated before /3/.

Massless superrenormalizable and (massless or massive) nonrenormalizable theories have in common that they cannot be constructed in perturbation theory. Both sorts of theories one may attempt to construct via regularization, i.e. introducing a small physical mass m, $\to 0$ in the limit, in the superrenormalizable (IR) case, and a cutoff Λ, $\to\infty$ in the limit, in the nonrenormalizable (UV) case. Restricting for simplicity the discussion to ϕ^4 theory, we are led to consider $(\phi^4)_{4-\varepsilon}$ theory, $\varepsilon=1,2$, and $(\phi^4)_{4+\varepsilon}$ theory, $\varepsilon=1,2,3$ ... . It is advantageous in both cases to let ε be generic, e.g. also complex, since for generic ε dimensional degeneracies are lifted that otherwise would lead to logarithms in most of the following formulae, and these logarithms are more complicated to analyse than the powers arising for generic dimension.

In the IR case /4/, in the Lagrangean

$$L = \frac{1}{2} \partial_\mu \phi_B \partial^u \phi_B - \frac{1}{4!} g_B \phi_B^4 - \frac{1}{2} m_B^2 \phi^2 \qquad (1)$$

m_B^2 can (at least in perturbation theory) be adjusted such that the physical mass is m. Since dim $g_B = \varepsilon$, the one dimensionless parameter in (1) is $g_B m^{-\varepsilon}$. The Fourier transforms of the vertex functions (VFs) (connected amputated one-particle-irreducible parts of the Green's functions) admit for nonexceptional momenta $p_1 \ldots p_{2n} \equiv (2n)$, the small-m expansion (with $k \leq \mathcal{L}$ for $\mathcal{L}$-loop graphs)

$$\Gamma_B((2n);\, m,\, g_B,\, \varepsilon) =$$
$$= p_{oo}((2n);\, g_B,\, \varepsilon) + \sum_{j=1}^{\infty} \sum_{k=o}^{\infty} m^{2j-\varepsilon k} p_{jk}((2n);\, g_B, \varepsilon) \qquad (2)$$

where p_{oo} has singularities at (in sufficiently high orders of perturbation theory, all) positive rational ε (as well as at nonpositive rational ε, not of interest here). These singularities are cancelled by terms in the double sum, identically in m, g_B, and the momenta, since the l.h.s. in (2) is (for Re $\varepsilon > o$) ε - singularity free.

Factorization properties of the p_{jk}, derivable from Wilson expansions, allow in (2) partial summations leading to

$$\Gamma_B((2n);\, m,\, g_B,\, \varepsilon) = p_{oo}((2n);\, g_B,\, \varepsilon) +$$
$$+ \sum_{j=1}^{\infty} g_B^{2j/\varepsilon} \sum_{L=1}^{L_j} C_{jl}\, (g_B\, m^{-\varepsilon}, \varepsilon)\, P_{jl}\, ((2n);\, g_B, \varepsilon) \qquad (3)$$

where the C_{jl} are expressible in terms of certain Green's functions of the massive theory. From the expressions for the lowest C_{jl} it can be seen that their ε- singularities are m-independent. A sufficient condition for the existence of the $m \to 0$ limit in (3) is

$$\lim_{m\to o} C_{jl}\ (g_B\ m^{-\varepsilon},\varepsilon) = C_{jl}\ (\varepsilon) \qquad (4)$$

where the C_{jl} (ε) have singularities such that the cancellation of ε-singularities on the r.h.s. of (3) persists in the limit. The success of the theory of critical phenomena /5/, for $\varepsilon = 1$ and $\varepsilon = 2$, based on this assumption (or assumptions equivalent hereto) lends strong support to (4). For $\varepsilon = 1$ and $\varepsilon = 2$, on the r.h.s of (3) there appear powers of $\ln g_B$ before, and after, the limit $m \to o$.

In the UV case, we choose the Lagrangean

$$L = \frac{1}{2}\,\partial_\mu\,\phi_B \prod_{r=1}^{R}\,(1+a_r^2\,\Lambda^{-2}\,\Box\,)\partial^u\,\phi_B - \\ - \frac{1}{4!}\,g_B\,\phi_B^4 - \frac{1}{2}\,m_{Bo}^2\,\phi_B^2 \qquad (5)$$

with dim $g_B = -\varepsilon$, where m_{Bo}^2 is the bare mass of the massless theory, to which we restrict ourselves for simplicity. (The kinetic term in (5) is written in symbolic fashion; for its more accurate form see /6/.) The VFs admit the large- Λ expansions (with $k \leq \mathcal{L}$ for $\mathcal{L}$-loop graphs)

$$\Gamma_{\Lambda B}\ ((2n);\ g_B,\varepsilon) = \sum_{j=o}^{\infty}\ \sum_{k=o}^{\infty}\ \Lambda^{-2j\ +\ \varepsilon k}\ f_{jk}\ ((2n);\ g_B,\varepsilon) \qquad (6)$$

with $f_{oo}((2n);\ g_B,\varepsilon) \equiv p_{oo}((2n);\ g_B,-\varepsilon)$ of (2). The factorization properties of the f_{jk} follow from the effective Lagrangean from which (6) can be generated,

$$L_\Lambda = \frac{1}{2}\,\partial_\mu\,\phi_B\,\partial^u\,\phi_B - \frac{1}{4!}\,g_B\,\phi_B^4 + \\ + \frac{1}{2}\,\partial_\mu\,\phi_B\,[\prod_{r=1}^{R}\,(1+a_r^2\,\Lambda^{-2}\Box)-1]\,\partial^u\,\phi_B + \\ + \sum_{\substack{r=o\\ r+s\geq 2}}^{\infty}\ \sum_{s=1}^{\infty}\ \sum_{\nu=1}^{rs}\ frs\nu\ (g_B\Lambda^{\varepsilon},\varepsilon)(\text{"}D^{2r}\phi_B^{2s}\text{"})_\nu\ g_B^{s-1}\ \Lambda^{4-2r-2s} \qquad (7)$$

Here $("D^{2r}\ \phi_B^{2s}")_\nu$ are the n_{rs} monomials, containing 2r derivatives and 2s factors ϕ_B, that are linearly independent at zero momentum transfer. The first two terms on the r.h.s. of (7) specify the Feynman rules, and all other terms are to be treated as (repeated) insertions into the VFs, with the rules of $4+\varepsilon$ - dimensional analytic integration /7/ being used. (Remark: In a similar way, one may generate the expansion (3) from an effective action which is, however, not the space-time integral of a local Lagrangean.)

The $j = 0$, $k > 0$ terms in (6), which have no couterparts in (3), are removed by "renormalization". Define functions

$$\beta(t,\varepsilon) = \varepsilon t + b_o(\varepsilon)\ t^2 + b_1\ (\varepsilon)\ t^3 + \ldots \tag{8a}$$

$$\gamma(t,\varepsilon) = c_o(\varepsilon)\ t^2 + c_1(\varepsilon)\ t^3 + \ldots \tag{8b}$$

from

$$[\partial/\partial p^2]\ \Gamma_{\Lambda B}\ (p(-p);\ g_B,\varepsilon)|_{p=o} = i\ \underline{Z}_3\ (g_B\ \Lambda^\varepsilon,\varepsilon)^{-1} \tag{9a}$$

$$\Gamma_{\Lambda B}\ (0000;\ g_B,\varepsilon) = -i\ \Lambda^{-\varepsilon} g\ (g_B\ \Lambda^\varepsilon,\varepsilon)\ \underline{Z}_3\ (g_B\ \Lambda^\varepsilon,\varepsilon)^{-2} \tag{9b}$$

where

$$\underline{Z}_3\ (t,\varepsilon) \equiv \exp[-\int_o^t ds\ \beta(s,\varepsilon)^{-1}\ \gamma(s,\varepsilon)]$$

and

$$g(t,\varepsilon) \equiv t\ \exp\{\varepsilon\int_o^t ds[\beta(s,\varepsilon)^{-1} - \varepsilon^{-1}\ s^{-1}]\}.$$

Then setting

$$\phi_B\ \underline{Z}_3\ (g_B\ \Lambda^\varepsilon,\varepsilon)^{-1/2} = \phi \tag{10a}$$

$$\Lambda^{-\varepsilon} g(g_B\ \Lambda^\varepsilon,\varepsilon) = g\mu^{-\varepsilon} \tag{10b}$$

(7) becomes /1/

$$L_\Lambda = \frac{1}{2} \partial_\mu \phi \, \partial^u \phi - \frac{1}{4!} g \mu^{-\varepsilon} \phi^4 + \sum_{\substack{r=o \\ r+s\geq 3}}^{\infty} \sum_{s=1}^{\infty} \sum_{\nu=1}^{n_{rs}} c'_{rs\nu} (g\mu^{-\varepsilon}\Lambda^{\varepsilon},\varepsilon) \, ("D^{2r}\phi^{2s}")_\nu \, (g\mu^{-\varepsilon})^{s-1} \Lambda^{4-2r-2s} \tag{11}$$

where $c'_{rs\nu}$ are power series in their first arguments with coefficients meromorphic in ε. Their singularities are such that in the large-Λ expansion of "renormalized" VFs

$$\Gamma_\Lambda ((2n);\, g\mu^{-\varepsilon},\varepsilon) = f_{oo}((2n);\, g\mu^{-\varepsilon},\varepsilon) + \sum_{j=1}^{\infty} \sum_{k=o}^{\infty} \Lambda^{-2j+\varepsilon k} h_{jk} ((2n);\, g\mu^{-\varepsilon},\varepsilon) \tag{12}$$

the ε -singularities (at rational ε) of f_{oo} in $0<\mathrm{Re}\,\varepsilon<4R$ are cancelled, identically in Λ, g, and the momenta, by terms in the double sum.

In view of (11), the expansion analogous to (3) has coefficients involving powers of the functions

$$c'_{rs\nu} (g\mu^{-\varepsilon}\Lambda^{\varepsilon},\varepsilon)(g\mu^{-\varepsilon}\Lambda^{\varepsilon})^{(4-2r-2s)/\varepsilon} \equiv \overline{C_{rs\nu}}(g\mu^{-\varepsilon}\Lambda^{\varepsilon},\varepsilon).$$

These functions can directly be expressed in terms of VFs Γ_Λ at zero momenta and derivatives there, e.g.

$$C_{o31} (g\mu^{-\varepsilon}\Lambda^{\varepsilon},\varepsilon) = -i\,(6!)^{-1}\,(g\mu^{-\varepsilon})^{-2-2/\varepsilon} \underset{\text{from Re } \varepsilon>2}{\text{anal. cont.}} \Gamma_\Lambda(000000;g\mu^{-\varepsilon},\varepsilon). \tag{13}$$

For the existence of the limits Γ_∞, as $\Lambda\to\infty$, of the Γ_Λ it is, in analogy to (4), sufficient that

$$\lim_{\Lambda\to\infty} C_{rs\nu} (g\mu^{-\varepsilon} \Lambda^{\varepsilon},\varepsilon) = C_{rs\nu}(\varepsilon) \tag{14}$$

exists; here the ε-singularities of the limitands are fixed implicitly by the conditions of ε-singularity cancellation in (12) identically in Λ as mentioned before, and are for the lowest $C_{rs\nu}$ Λ-independent as they are for (13). In this case, as well as in general, these singularities are IR singularities of the VFs appearing in the formulae.

The renormalization theory of Bogoliubov and Shirkov /8/ shows /1/ that in

$$\bar{L} = \frac{1}{2}\partial_\mu \phi \partial^\mu \phi - \frac{1}{4!} g\mu^{-\varepsilon}\phi^4 + \sum_{\substack{r=0\\ r+s\geq 3}}^{\infty}\sum_{s=1}^{\infty}\sum_{\nu=1}^{n_{rs}} \bar{C}_{rs\nu}(\varepsilon)\,(\text{"}D^{2r}\phi^{2s}\text{"})_\nu\,(g\mu^{-\varepsilon})^{s-1+(2r+2s-4)/\varepsilon} \tag{15}$$

the coefficients $\bar{C}_{rs\nu}(\varepsilon)$ can be chosen such that in the corresponding expansions of the VFs no ε-singularities for Re $\varepsilon > 0$ arise. It follows that, if $\Gamma_\infty = \lim_{\Lambda\to\infty} \Gamma_\Lambda$ exist in the sense explained before $C_{rs\nu}(\varepsilon)$ are for $0 <$ Re $\varepsilon < R$ a family of coefficients $\bar{C}_{rs\nu}(\varepsilon)$ as described. The formulae, like (13), expressing the $C_{rs\nu}(\varepsilon)$ in terms of Γ_∞ show that existence of the $C_{rs\nu}(\varepsilon)$ is not only sufficient but also necessary for the Γ_∞ to exist, at least for generic ε.

That the Γ_∞ computed from (15) with the particular choice $C_{rs\nu}(\varepsilon)$ for the $\bar{C}_{rs\nu}(\varepsilon)$ have a greater chance to describe a unitary theory (causality, in the sense of Bogoliubov and Shirkov /8/, being a consequence of the local form of (15)) than less constrained ones is suggested by the form the differential-vertex-operation (DVO) relations /9/ for $n\,\Gamma_\Lambda((2n);\, g\mu^{-\varepsilon},\varepsilon)$ and $[\partial/\partial g]\Gamma_\Lambda\,((2n);\, g\mu^{-\varepsilon},\varepsilon)$ take in the limit $\Lambda\to\infty$: comparison with the DVO relation for $\Lambda[\partial/\partial\Lambda]\;\Gamma_\Lambda\,((2n);\, g\mu^{-\varepsilon},\varepsilon)$ shows that $n\,\Gamma_\infty$ and $[\partial/\partial g]\;\Gamma_\infty$ will be represented by the sum of the two VFs with only the (renormalized) operator insertions $\partial_\mu\phi\partial^\mu\phi$ and ϕ^4, at zero momentum, characteristic for the unregularized Lagrangean. Similarly, the assumption that the Γ_∞ exist for a range of ratios of the a_r in (5) leads to $[\partial/\partial a_r]\ \Gamma_\infty = 0$ in this range, upon examing the DVO relations for $[\partial/\partial a_r]\ \Gamma_\Lambda$.

The expansion deriving from (15) (with $C_{rs\nu}(\varepsilon)$ for $\bar{C}_{rs\nu}(\varepsilon)$) is not suitable for computing the Γ_∞ for momenta $\gtrsim \mu g^{-1/\varepsilon}$. A resummation of that expansion will then be required, and it has presumably to be based on conformal invariance, with anomalous dimensions, at large momenta. In fact, if $\beta(t,\varepsilon)$ of (8a) has, as function of t, a first positive zero at $t_\infty(\varepsilon)$, then the Γ_∞ are expected to be scale invariant at large momenta with dimension $1 + \frac{1}{2}\varepsilon + \gamma(t_\infty(\varepsilon),\varepsilon)$ of the ϕ-operator, and also to be conformal invariant /9/. The analogy of the problems encountered in massless superrenormalizable and nonrenormalizable theories was first observed by Parisi /10/, and he has proposed approximation schemes /11/ that in effect deal with the resummation problem which here remains still unsolved.

There is till now no independent evidence that the limit $\Lambda\to\infty$ for the Γ_Λ does exist for $\varepsilon=1,2$, or generic, in contrast to the IR case discussed before. Only very simple approximations to functions in nonrenormalizable theory can be studied so far, an interesting example, with encouraging result, being /12/.

Final remark: If one uses a more general regularization than the specific one in (5), e.g. employing a sharp cutoff in momentum space or a lattice in configuration space, the expansion of $\Gamma_{\Lambda B}$ for large Λ will not take the form (6), and in particular a local effective Lagrangean of the form (7) will not exist. This makes it implausible that a causal and unitary (and, in case of a noncovariant cutoff, covariant) limit theory can then be obtained by letting $\Lambda\to\infty$ for "renormalized" Γ_Λ if $\varepsilon>0$. This contrasts with the case $\varepsilon=0$, where the limit theory (upon replacing the definitions (10) by more conventional ones then appropriate, or as in /13/ for the massive case) is (in perturbation theory at least) always the usual renormalized one, even with noncovariant cutoff.

References

/1/ K.SYMANZIK, Comm. math. Phys. (DESY 75/12)

/2/ K.SYMANZIK, DESY 75/24

/3/ K.SYMANZIK, in: Lecture Notes in Physics, Vol. 39, Ed. H. ARAKI (Kyoto Symposium 1975)

/4/ K.SYMANZIK, Lett. Nuovo Cimento 8, 771 (1973); in: 1973 Cargèse Lectures in Physics, Ed. E. Brézin (DESY 73/58)

/5/ E.BRÉZIN, J.C.LE GUILLOU, J.ZINN-JUSTIN, D Ph-T/ 74 / 100, Saclay, Dec. 1974

/6/ A.PAIS, G.UHLENBECK, Phys. Rev. 79, 145 (1950)

/7/ K.G.WILSON: Phys. Rev D7, 2911 (1973);G.'t HOOFT, M.Veltman, in: Particle Interactions at Very High Energies, Part B, Ed. D.SPEISER et al., Plenum Press, New York 1974

/8/ N.N.BOGOLIUBOV, D.V.SHIRKOV: Introduction to the Theory of Quantized Fields. Interscience Publ.,New York 1959

/9/ B.SCHROER, Lett. Nuovo Cimento 2, 867 (1971)

/10/ G.PARISI, Cargèse Lectures 1973, Columbia Univ. preprint

/11/ G.PARISI, Note Interne 573 (1974), 621 (1975), Università di Roma

/12/ B.A.ARBUZOV, A.T.FILIPPOV, Nuovo Cimento 38, 796 (1965)

/13/ R.SCHRADER, Colloqu. on Math. Methods in QFT, Marseille, June 1975

Acta Physica Austriaca, Suppl. XVI, 185–200 (1976)

Challenging Problems in Singular Dynamical Systems

by

John R. Klauder
Bell Laboratories
Murray Hill, New Jersey 07974

Abstract

Fundamental problems common to many systems with singular dynamics are qualititively discussed. In particular, discontinuous perturbations in quantum mechanics, field theory and statistical mechanics are presented. In addition, the solution of a class of covariant nonrenormalizable field theories is outlined.

Models involving singular phenomena arise in a variety of dynamical systems: simple particle problems; critical behavior in statistical mechanics; field theory models, etc. In turn, the singular behavior manifests itself in nonanalytic terms in perturbation expansions, characteristic divergences of correlation phenomena, need for divergent renormalization counterterms, etc. And even as the diversity of systems in which interesting singular behavior enlarges, there is a growing awareness of the general equivalence of certain seemingly disparate problems. Such general equivalences by now interconnect various topics in field theory, statistical mechanics, percolation theory, stochastic processes, etc. Phase transitions, for example, are recognized to apply, at least potentially, to virtually all systems having infinitely many degrees of freedom.

While there are numerous problems ready for straightforward attack--often made so in part by the multitude of cross fertilizations now extant--we wish to comment on some general conceptual problems that we feel are fundamental and still unanswered, at least in any fully satisfactory way. Of course, different researchers may perceive other basic issues as being fundamentally important, and this variation may have its origin in different degrees of satisfaction (or dissatisfaction) with the present state of the art.

Motivation

Much of our discussion is motivated by a recently emphasized distinction between continuous and discontinuous perturbations.[1] In its simplest manifestation, namely in the context of conventional single-particle quantum mechanics, examples such as $P^2 + Q^2 + \lambda V(Q)$ where $V(Q) = Q, Q^2, Q^4$, or $\exp Q^4$, etc., represent continuous perturbations in as much as when $\lambda \downarrow 0$ the Hamiltonian goes to the harmonic oscillator Hamiltonian $P^2 + Q^2$ (strong resolvent convergence). More intuitively the energy eigenvalues all continuously return to those of the harmonic oscillator in spite of the fact that their dependence on λ (entire, analytic in a circle, asymptotic, nondifferentiable, etc.) are entirely different. On the other hand, examples such as $P^2 + Q^2 + \lambda V(Q)$ where $V(Q) = |Q|^{-3}$, $|Q-c|^{-\alpha}$, $\alpha > 2$, $\exp(1/Q^2)$, $\sec^4(Q)$, etc. represent discontinuous perturbations inasmuch as when $\lambda\downarrow 0$ the Hamiltonian does not pass to the harmonic oscillator Hamiltonian, but does pass continuously to another, a "pseudofree" Hamiltonian with generally different eigenvalues and eigenfunctions. Relative to the pseudofree Hamiltonian the perturbation in question is a continuous perturbation and may have all the various forms of λ dependence alluded to above. In this simplest class of examples the nature of the pseudofree model is fairly evident being characterized by the introduction of Dirichlet boundary conditions in the free Hamiltonian at the point(s) of singularity in the perturbation.[2]

Abstraction

The fundamental question that naturally arises is how general are such phenomena, and where and how might they arise in systems of infinite degrees of freedom. It is this author's conviction that such phenomena may arise in the context of highly singular quantum field theories, especially those presently classified as nonrenormalizable.

Hard Core Picture

To make this view more plausible consider the following heuristic argument.[1,3] A functional integral formulation of Euclidean quantum theory may be formally expressed by the integral

$$E(h) = \mathcal{N}\int e^{i\int h\Phi d^n x - W_0(\Phi) - \lambda W_I(\Phi)} \mathcal{D}\Phi$$

where $\mathcal{N}$ is a normalization factor, $x \in R^n$, n space-time dimension, $\Phi = \Phi(x)$ denotes a random field, $h = h(x)$ a test function, and $W_0(\Phi)$ and $W_I(\Phi)$ the free and interacting actions, respectively. Loosely speaking, all random fields Φ for which $W_I(\Phi) = \infty$ and $W_0(\Phi) < \infty$ are projected out of the integrand - - as with a "hard core" - - for all $\lambda > 0$. If the measure of such fields, in some general sense, is nonzero, then what emerges as $\lambda\downarrow 0$ is not the descriptor of the free field but rather of a pseudofree field which clearly contains the vestiges of the hard core interaction. In this sense we see how the interaction acts as a discontinuous perturbation relative to the free theory.

Field Theory Examples

As yet no covariant quantum field theory is known that exhibits such behavior, but one highly idealized example exists that clearly and cleanly illustrates the desired phenomena.[4] This model is the so-called independent-value model, a name derived from its highly specialized kinematics. For this example
$W_o(\Phi) = \frac{1}{2}m^2\int\Phi^2 d^n x$ and $W_I(\Phi) = \int|\Phi|^p d^n x$, $p > 2$.
Since there are no derivatives of the field, the field has independent values at every space-time point and as such the characteristic functional necessarily has the form

$$E(h) = \exp\{-\int L[h(x)]d^n x\}$$

for some even function $L[h]$. This is just the form leading to infinitely-divisible characteristic functions[5] and general arguments assert that

$$L[h] = ah^2 + \int_{|u|>o}[1 - \cos(uh)]d\sigma(u)$$

where $a \geq o$ and $\sigma(u)$, $u \in R$, is a nondecreasing measure such that

$$\int_{|u|>o} u^2/(1+u^2)d\sigma(u) < \infty$$

The free theories are evidently characterized by $\sigma \equiv o$ and $a=\frac{1}{2}m^{-2}$, and it may be determined[4] that a proper choice for any interacting model $(p > 2)$ is given by $a \equiv o$ and

$$d\sigma(u) = |u|^{-1}\exp(-\tfrac{1}{2}m^2u^2 - \lambda|u|^p)du$$

Evidently, in every such case as $\lambda\downarrow 0$, the interacting theories do not return to any free theory but to a unique pseudofree (PF) theory for which $a \equiv o$ and $d\sigma_{PF}(u) = |u|^{-1}\exp(-\frac{1}{2}m^2u^2)du$. Since this example is treated elsewhere we do not pursue it further here (although this solution will arise in other guises in our later discussion).

Unapproachability Via Lattice Cutoffs

Presumably the quoted solution for the independent-value model could be approached by the use of an ultraviolet cutoff in the form of a space-time lattice. In such a case one is led to consider products (for different lattice points and different lattice-point test function values) of an expression of the form

$$N\int e^{ih\Phi - \frac{1}{2}M^2\Phi^2 - \lambda_o|\Phi|^p} d\Phi$$

where N is a normalization factor, and M^2 (possibly negative) and λ_o are parameters related to m^2 and λ including also the lattice spacing. It suffices to examine $p = 4$, and to remark that, for the lattice form of the problem, the truncated four-point function is always nonpositive.[6] On the other hand, it follows directly from the general form for infinitely-divisible characteristic functions, i.e., for any a and nondecreasing σ, that the truncated four-point function is nonnegative. Consequently, there is absolutely no hope of approaching the no-cutoff, symmetry-determined solution by way of taking limits of the lattice-cutoff solutions!

This view has now been verified in several ways in the works of Caianiello and Scarpetta,[7] Kainz,[8] and through the recent comparative study by Marinaro[9] who has studied the branching equations (differential equations relating various Green's functions) for the cutoff and no-cutoff solutions pinpointing the inevitable distinctions in the two approachs.

Qualitative Discussion

Unphysical as the preceding example may be, it is nonetheless mathematically relevant as a test of the basic concepts. Clearly there are infinitely many fields for which $\Phi \in L^2$ but $\Phi \notin L^p$, $p > 2$; and this is simply not the case for a mass perturbation where $p = 2$. Unfortunately, this distinction is essentially heuristic since even the free measure is known not to be carried by L^2 but necessarily requires a larger space.[5] Indeed, the set of random fields with finite square integral has zero measure for the free theory.

We need a somewhat more refined description of suitable carrier spaces for the measures, and the following remarks hopefully point in a helpful direction. We know that the free measure remains equivalent to itself under translation of the carrier space by an element g if and only if $g \in L^2$.[10] On the other hand, for the interacting models and the solutions we have presented, the interacting measures become mutually singular under translation of the carrier space by <u>any</u> $g \neq 0$.[11] There is a genuine distinction between the support properties in this sense. Clearly further development of such characterizations is important to further understanding.

One additional point is worth discussing. Suppose we choose $W_I(\Phi) = \frac{1}{2}\int |\nabla\Phi|^2 d^n x$ instead of our earlier choice. Then superficially our basic criteria is again satisfied since there are many Φ for which $W_I(\Phi) = \infty$ and $W_0(\Phi) < \infty$. On the other hand, this case clearly represents a continuous perturbation of the free problem, and this has its trace in the nature of the carrier spaces for the measure. As indicated above, the carrier space of the free measure is invariant under translation by any $g \in L^2$, and the carrier space of the (present) interacting theory is also invariant under translation by any g in a dense subset of L^2 completey unlike the carrier space for the other nonlinear interactions.

The distinction in carrier spaces in the former case is deep-seated and fundamental, while the distinction in the latter case is relatively minor and quantitative. Crudely speaking, in the former case the change of support occurs "everywhere", while in the latter case it seems to be only at the "edges"; in the former case the support is descriptively pictured as "Swiss cheese", riddled with hard-core generated holes in the support, while in the latter case only the "edge" is affected on an otherwise solid block. This basic topological-type distinction appears to be crucial in distinguishing certain continuous and discontinuous perturbations from each other. Clearly a "Swiss cheese" picture relative to the free support may turn into an "edge" picture relative to the appropriate pseudofree support paralleling the change of a discontinuous perturbation into a continuous one relative to the appropriate pseudofree theory.

Relevance for Covariant Models

When it comes to discuss covariant models, we can only argue by analogy. For a covariant model $W_o(\Phi) = \frac{1}{2}\int[(\nabla\Phi)^2 + m^2\Phi^2]d^nx$ and $W_I(\Phi) = \int|\Phi|^p d^nx$. Sobolev-type inequalities ensure that $W_o(\Phi) < \infty$ implies that $W_I(\Phi) < \infty$ whenever $p \leq 2n/(n-2)$, and decidedly not whenever $p > 2n/(n-2)$.[1,12] Mathematically, it should make no real difference in the basic structure whether we deal with one or another quadratic free theory, namely the covariant case with gradients or the independent-value case without gradients. Consequently, one is led to conjecture the hard core picture, qualitatively as we have already seen for the independent-value models, whenever $p > 2n/(n-2)$ -- and this is precisely the category of scalar field theories characterized as nonrenormalizable in the context of renormalized perturbation theory. Thus, it is plausible that nontrivial covariant nonrenormalizable quantum field theories may be understood in the general framework of discontinuous perturbations.

Soluble Covariant Nonrenormalizable Model

In fact there does exist a soluble covariant nonrenormalizable model but one that is essentially trivial and which does not really correspond to a discontinuous perturbation.[13] The model in question pertains to an Abelian gauge vector field $\phi_\mu(x)$ in a two-dimensional space time with action functionals given by

$$I = \int\{j^\mu\phi_\mu + (-\tfrac{1}{4}f_{\mu\nu}f^{\mu\nu}) + \lambda\mathcal{L}'[-\tfrac{1}{4}f_{\mu\nu}f^{\mu\nu}]\}d^2x$$

where $f_{\mu\nu} = \partial_\mu\phi_\nu - \partial_\nu\phi_\mu$, $\mathcal{L}'$ is nonlinear and j^μ represents a conserved external current (and thus $j^\mu = \varepsilon^{\mu\nu}\partial_\nu h$, for some h). Since the interaction involves "derivative coupling" such models are nonrenormalizable even in two dimensions. The action is invariant under the gauge transformation $\phi_\mu \to \phi_\mu + \partial_\mu\Lambda$ and this fact may be exploited to find a solution.

In the Coulomb gauge, where $\phi_1 \equiv 0$, the only field component $f_{01} = -\partial_1\phi_0$, and the Euler-Lagrange equation of motion is strictly a constraint since no time derivatives appear in the problem. Elimination of the constraint yields $I = -\int\mathcal{H}[h]d^2x$, where $\mathcal{H}$ denotes the Hamiltonian derived from the indicated Lagrangian. Consequently, determination of the standard Green's function generating functional

$$Z\{j\} = \mathcal{N} \int e^{iI} \mathcal{D}\phi_\mu$$

which entails an integration over fields modulo gauge transformations, is necessarily trivial and given by

$$Z\{j\} = e^{-i\int \mathcal{H}[h] d^2x}$$

since after elimination of the constraint there are no integration variables remaining.

This answer must also emerge if the calculation is performed in another gauge, say, a gauge in which $\phi_0 \equiv 0$ and $f_{01} = \partial_0\phi_1 \equiv \dot{\phi}_1$. Functional integrations involving general functions of $\dot{\phi}_1$ are properly formulated in the phase-space formulation, and in that framework it readily follows that the previously determined answer emerges.[13]

Operator Structure

Most significant is the operator realization that is required for the free and nonfree models of this type.[13] Introducing a small mass μ in the free case to remove the infrared divergence readily leads to a relevant operator realization given by

$$\phi_1(t,w) = \frac{1}{\sqrt{2\mu}} [e^{-i\mu t}A(w) + A^\dagger(w)e^{i\mu t}]$$

when $x = (x^0,x^1) \equiv (t,w)$, and

$$[A(w),A^\dagger(w')] = \delta(w-w')$$

and $A(w)|0\rangle = 0$. It follows quite directly that

$$\langle 0|T \exp(i\int h\dot{\phi}_1 d^2x)|0\rangle \xrightarrow[\mu\to 0]{} \exp(-i\tfrac{1}{2}\int h^2 d^2x)$$

which is therefore a realization of the free model as a limit of regularized operator-based expressions.

For every nonfree model a radically different operator realization or construction is required which is conveniently given by

$$\phi_1(t,\omega) = \int A^\dagger(w,\xi)e^{i\mu\hbar t}\xi\ e^{-i\mu\hbar t}A(w,\xi)d\xi$$

$$+ \frac{1}{\sqrt{\mu}} \int [A^\dagger(w,\xi)e^{i\mu\hbar t}\xi u(\xi)+u(\xi)\xi e^{-i\mu\hbar t}A(w,\xi)]d\xi$$

where $\xi \in R, \hbar = \hbar(-i\partial/\partial\xi,\xi)$ and $u(\xi)$ (real and even) are to be further specified. Here the basic operators fulfill

$$[A(w,\xi)\ ,\ A^\dagger(w',\xi')] = \delta(w-w')\delta(\xi-\xi')$$

and $A(w,\xi)|0> = 0$. A straightforward, but somewhat lengthy, calculation shows that

$$<0|T\ \exp(i\int \dot{h}\phi_1 d^2x)|0> \underset{\mu\downarrow 0}{\longrightarrow} \exp\{-i\int \mathcal{H}[h]d^2x\}$$

where

$$\mathcal{H}[h] \equiv \int u(\xi)(e^{-ih\xi}-1)\hbar(-i\partial/\partial\xi,\xi)(e^{ih\xi}-1)u(\xi)d\xi$$

and it is not difficult to choose $\hbar$ and u to correspondent to any desired pregiven Hamiltonian $\mathcal{H}$. It should be remarked that the basic bilinear operator form of the interacting solution is dictated by general theorems and is by no means arbitrary.[14]

Additional Comments

The essential fact to observe about this model is that while the infrared regularized free model can be realized by a linear expression in $A(w)$ and $A^\dagger(w)$, every infrared regularized interacting model requires (the equivalent of) a bilinear construction and an auxiliary variable ξ in the Fock representation of the basic operators. There is no escape from this dichotomy of construction. Moreover, if with $\mu > 0$, we consider $\lambda\downarrow 0$, then we end up with a pseudofree field, a field operator bilinear in $A^\dagger$ and A, and still involving the variable ξ, and moreover one to which all of the interacting examples are continuously connected. In the limit $\mu\downarrow 0$, however, the pseudofree field Green's function generating functional becomes $\exp(-i\frac{1}{2}\int h^2 d^2x)$ exactly as for the free field. Thus, although the interaction indeed represents a discontinuous perturbation of the infrared regularized models, this discontinuity disappears as the regularization is removed.

In view of the ultimate triviality of such models (only a constraint in one gauge; just a phase factor for the Green's function generating functional) it is perhaps not too surprising that a genuine discontinuous perturbation does not survive when the regularization parameter μ vanishes.

Ultraviolet and Infrared

Divergence difficulties in field theory may occur not only at high momenta (ultraviolet) but at low momenta (infrared) as well. One school of thought asserts that traditional problems encountered in ultraviolet divergences are rather similar to those encountered in associated infrared divergences.[15] A suggestive argument first observes that the propagator for a scalar particle (of mass m in n dimensions) cutoff at momentum Λ, enters schematically as

$$\int^{\Lambda} \ldots \frac{1}{k^2+m^2} \ldots d^n k$$

in perturbation calculations. Rescaling according to $k \to k\Lambda$ this expression becomes

$$\int^{1} \ldots \frac{\Lambda^{n-2}}{k^2+m^2/\Lambda^2} \ldots d^n k$$

which shows the relationship of an ultraviolet limit $\Lambda \to \infty$ with a critical behavior $m^2/\Lambda^2 \to 0$ in an alternative system of fixed ultraviolet cutoff and with certain coupling constant modifications. But certainly such an argument is oversimplified on at least two counts.

Firstly, the traditional problems one deals with involve short-range forces (in x space), which for nonquadratic interactions lead to long-range forces (in k space). Take as an example a field theory model for a complex field with Euclidean action

$$I = -\int\{\tfrac{1}{2}|\nabla\Phi(x)|^2 + \tfrac{1}{2}m^2|\Phi(x)|^2 + \lambda|\Phi(x)|^p\}d^n x$$

Scaling arguments suggest that if $p > 2n/(n-2)$ (e.g., if $n = 4$, then $p > 4$), the last term makes no contribution compared to the first term in an infrared scaling limit, while if $p < 2n/(n-2)$ nonclassical infrared scaling behavior exists. In addition, analogous scaling arguments suggest that it is precisely when $p > 2n/(n-2)$ that the ultraviolet divergences become unmanageable and one deals with a nonrenormalizable theory, while for $p < 2n/(n-2)$ the theory is superrenormalizable. Since irrelevant infrared scaling is different in detail from superrenormalizability, there is no reason to expect nonclassical infrared scaling to correspond to nonrenormalizability.

Secondly, one can concoct an associated mathematical model in which the roles of k and x are interchanged leading to a model with Euclidean action

$$\tilde{I} = - \int\{\tfrac{1}{2}|\nabla_k\tilde{\Phi}(k)|^2 + \tfrac{1}{2}m^2|\tilde{\Phi}(k)|^2 + \lambda|\tilde{\Phi}(k)|^p\}d^nk$$

Here the conventional meanings of infrared and ultraviolet have been interchanged, and now one observes that if $p > 2n/(n-2)$, the model has nonrenormalizable infrared behavior and superrenormalizable ultraviolet behavior. If $p = 4$, for example, one notes that

$$\tilde{I} = - \int\tfrac{1}{2}(x^2+m^2)|\Phi(x)|^2d^nx$$

$$+ \lambda(2\pi)^n\int\Phi^*(x_1)\Phi^*(x_2)\Phi(x_3)\Phi(x_4)\delta(x_1+x_2-x_3-x_4)d^nx_1\ldots d^nx_4$$

which makes evident the source of the infrared trouble as long-range nonlinear forces. The present model renders nonrenormalizable ultraviolet behavior more visible in the familiar large distance realm.

Lattice Space Nonrenormalizability

A lattice space version of the preceding model makes it appear as a statistical mechanics problem. With unit lattice spacing one considers a classical Hamiltonian (ignoring inessential factors) of the form

$$\tilde{H} = \Sigma\tfrac{1}{2}(|r|^2+m^2)|\Phi_r|^2 + \lambda\Sigma\Phi^*_{r_1}\Phi^*_{r_2}\Phi_{r_3}\Phi_{r_4}\delta_{r_1+r_2,r_3+r_4}$$

where $r = (r^1,r^2,\ldots,r^n)$, $|r|^2 = \Sigma(r^j)^2$ and $r^j = \ldots,-2,-1,0,1,\ldots$ denote the lattice coordinates. In this context one faces the problem of the thermodynamic limit which in view of the long-range forces is far from trivial. Indeed it may be argued that if the ultraviolet cutoff cannot be removed in the original field theory problem, then the thermodynamic limit for the transformed lattice problem does not exist.

Whether or not the thermodynamic limit exists is evidently influenced by the term $\Sigma|r|^2|\Phi_r|^2$ which puts certain limits on the magnitude of spin deviations at distant lattice sites. Suppose we arbitrarily replace that term by $\Sigma|r|^{2\xi}|\Phi_r|^2$ when $\xi \geq 0$ is a parameter at our disposal. If $\xi > n/2$, there are sufficient limitations so that the thermodynamic limit exists. If (for the quartic interaction under discussion) $n/2 \geq \xi \geq n/4$, renormalizations are needed to ensure that a conventional thermodynamic limit exists, while if $\xi < n/4$ no presently known techniques apply. Evidently the worst of such cases arises when $\xi = 0$, and in fact in this case it matters little what n value is chosen. Yet when $\xi = 0$ we can appeal to the solution presented earlier for the independent-value field theory model.

Soluble Model

Consider the model described by the classical Hamiltonian

$$\tilde{H} = \frac{1}{2}\Sigma m^2|\Phi_r|^2 + \lambda\Sigma\Phi^*_{r_1}\Phi^*_{r_2}\Phi_{r_3}\Phi_{r_4}\delta_{r_1+r_2,r_3+r_4}$$

which corresponds to the previous case with $\xi = 0$ (and the resultant constant absorbed into a redefined m). For simplicity let n = 1 and interpret r as a lattice coordinate in one dimension. In calculating the statistical mechanics for such a system one is naturally inclined to truncate the sum (put the system in a box) and take the thermodynamic limit as a final step. Resist that temptation! Instead introduce the Fourier variable

$$\tilde{\Phi}(k) = (2\pi)^{-1/2}\Sigma e^{-ikr}\Phi_r$$

with inverse

$$\Phi_r = (2\pi)^{-1/2}\int_{-\pi}^{\pi} e^{ikr}\tilde{\Phi}(k)dk$$

in terms of which

$$\tilde{H} = \int_{-\pi}^{\pi}\{\frac{1}{2}m^2|\tilde{\Phi}(k)|^2 + (\lambda/2\pi)|\tilde{\Phi}(k)|^4\}dk$$

In this form the connection with independent-value models is evident, and based on the solution (or a simple generalization thereof) presented earlier, it is reasonably clear that

$$\mathcal{N}\int e^{\int_{-\pi}^{\pi}\{i\tilde{h}^*(k)\Phi(k) - \frac{1}{2}m^2|\Phi(k)|^2 - (\lambda/2\pi)|\Phi(k)|^4\}dk}\mathcal{D}\Phi$$

$$= \exp - \int_{-\pi}^{\pi} dk \int\{1 - \cos[\tilde{h}^*(k)z]\}\ e^{-1/2m^2|z|^2-(\lambda/2\pi)|z|^4}\ \frac{d^2z}{|z|^2}$$

$$= \exp\ -2\pi\int_{-\pi}^{\pi} dk\int_0^{\infty}\{1 - J_o[|\tilde{h}(k)|u]\}\ e^{-1/2m^2u^2-(\lambda/2\pi)u^4}\ \frac{du}{u}$$

An approach to this problem by taking the conventional thermodynamic limit is equivalent to approaching the field theory problem by introducing a cutoff that softens the "local" product $|\Phi(k)|^4$. Here the local product is in k space while previously it was in x space, but that makes no difference mathematically. Earlier we had argued that introduction of a cutoff (at least in the form of a space-time lattice) was doomed to failure. By analogy, analysis of the present long-range lattice spin model by conventional thermodynamic limits is likewise doomed to failure. Evidently what one needs to treat such problems as these are novel and unconventional views toward thermodynamic limits. It may conceivably be in this context that the mysteries of non-renormalizable interactions are unravelled.

Summary and Suggested Problems

We have attempted to provide intuitive arguments and some examples to suggest that singular dynamical systems constitute an area of great interest and potential. Many problems remain both at the conceptual and technical level. Among such problems we might mention the need for further examples and additional characterization of continuous and discontinuous perturbations. Take the simple dynamical system with Hamiltonian

$$H = \Sigma(P_n^2 + Q_n^2) + \lambda\Sigma|Q_n|$$

Construction of the appropriate quantum system as a direct product representation of the canonical commutation relations is straightforward. This example is clearly a continuous perturbation (even though $\Sigma Q_n^2 < \infty$ does not imply $\Sigma|Q_n| < \infty$, which is clearly an "edge" effect.) On the other hand, consider

$$H = \Sigma(P_n^2 + Q_n^2) + \lambda(\Sigma|Q_n|)^4$$

or even

$$H = \Sigma(P_n^2 + Q_n^2) + \lambda(\Sigma|Q_n|)^2$$

Are these examples also continuous perturbations? Possibly not, but that would be interesting to prove.

Many open problems with regard to the two-dimensional Abelian gauge field remain. A manifestly covariant operator analysis even at the cost of an indefinite metric would be very helpful in understanding such models. Addition of spinor dynamics especially when nonzero charge sectors are allowed could prove fascinating. Generalization to an internal symmetry having O(N) symmetry and a study of the $N \to \infty$ limit should prove instructive. Finally generalization of this class of models to non-Abelian, Yang-Mills type gauge groups should be attempted.

Some major questions with regard to measures on random fields naturally arise. How is one to characterize carrier spaces in some less than complete fashion so as to quantify the heuristic concepts of hard cores and discontinuous perturbations? Can some uniform dissimilarity of sets of zero measure be used to establish in a sensible way that the measures in question are weakly disconnected?

We have occasionally but briefly touched on the notion of pseudofree fields, namely the nonfree fields to which the interacting fields weakly pass as the coupling is turned off. Can such models be systematically studied without needing first to solve the whole problem? Consider the general generating functional represented by

$$< e^{\Phi(h)} > \equiv \frac{\int e^{b\Phi(h) - W(b\Phi)} \mathcal{D}\Phi}{\int e^{-W(b\Phi)} \mathcal{D}\Phi}$$

where $b > 0$ is an arbitrary parameter on the right side on which the result does not depend. Evaluating the derivative of this expression with respect to b at $b = 1$ leads to the constraint relation

$$< \Phi(h)e^{\Phi(h)} > = < \mathcal{W}e^{\Phi(h)} >$$

where (as a form)

$$\mathcal{W} = (d/db)\left[W(b\Phi) - <W(b\Phi)>\right]\Big|_{b=1}$$

For a pseudofree (or free) field, W is <u>quadratic</u>, i.e.,

$$W(b\Phi) \equiv \tfrac{1}{2}Q(b\Phi) = \tfrac{1}{2}b^2 Q(\Phi)$$

and thus

$$\mathcal{W} = Q(\Phi) - < Q(\Phi) >$$

More to the point, as a quadratic expression

$$\mathcal{W} = \int K(x,y)\left[\Phi(x)\Phi(y) - < \Phi(x)\Phi(y)>\right]d^n x d^n y$$

for some formal distribution K. One solution of the constraint relation is to choose a mean zero Gaussian field with

$$< \Phi(x)\Phi(y) > = K^{-1}(x,y)$$

but there potentially exist many <u>non</u>-Gaussian solutions and their discovery is of fundamental importance.

Lastly let us re-emphasize the view of superrenormalizable and nonrenormalizable models that arise in lattice examples. Superrenormalizable models correspond to thermodynamic limits of lattice systems with short-range forces, while nonrenormalizable models correspond to thermodynamic limits of lattice systems with long-range nonlinear forces as made clear in the models involving an x - k interchange. It is almost a contradiction of terms to ask for the thermodynamic limit of systems with long-range forces; but it is in the resolution of just such seemingly incompatible concepts that progress in overall understanding is sure to emerge.

References

1 J. R. KLAUDER, Acta Physica Austriaca Suppl. XI, 341 (1973); Phys. Letters 47B, 523 (1973).

2 B. SIMON, J. Funct. Anal. 14, 295 (1973); B. DeFacio and C. L. Hammer, J. Math. Phys. 15, 1071 (1974).

3 J. R. KLAUDER, "Soluble Models and the Meaning of Nonrenormalizability", Lecture Notes in Physics, Vol. 39 (Springer-Verlag, New York, 1975), p. 160.

4 J. R. KLAUDER, Acta Physica Austriaca 41, 237 (1975).

5 I. M. GEL'FAND and N. YA. VILENKIN, Generalized Functions, Vol. 4: Applications of Harmonic Analysis, translated by A. Feinstein (Academic Press, 1964).

6 J. LEBOWITZ, Commun. Math. Phys. 35, 87 (1974); B. Simon and R. Griffiths, Commun. Math. Phys. 33, 145 (1973).

7 E. R. CAIANIELLO and G. SCARPETTA, Nuovo Cimento 22A, 448 (1974); Nuovo Cimento Letters 11, 283 (1974).

8 W. KAINZ, Lett. Nuovo Cimento 12, 217 (1975).

9 M. MARINARO, "Comparison of Branching Equations for a Non-renormalizable Model", Bell Labs - University of Salerno, preprint (to be published).

10 S. KAKUTANI, Ann. Math. 49, 214 (1948); L. A. Shepp, Ann. Math. Statistics 36, 1107 (1965).

11 G. C. HEGERFELDT and J. R. KLAUDER, Nuovo Cimento 10A,723 (1972).

12 O. A. LADYZENSKAJA, V. A. SOLONNIKOV and N. N. URAL'CEVA, Linear and Quasi-Linear Equations of Parabolic Type, Trans. of Math. Mono., Vol. 23 (American Math. Society, 1968).

13 J. R. KLAUDER, Phys. Rev. D 12, 1590 (1975).

14 J. R. KLAUDER, Phys. Rev. Lett. 28, 769 (1972); Lectures in Theoretical Physics, Vol. XIVB, W. Britten, ed. (Colorado Associated University Press, 1974), p. 329; M. B. Ruskai and J. R. Klauder, J. Math. Phys. 14, 1199 (1973).

15 See, e.g., K. G. WILSON and J. B. KOGUT, Phys. Rep. 12C, 75 (1974); K. G. Wilson, Rev. Mod. Physics 47, 773 (1975).

Acta Physica Austriaca, Suppl. XVI, 201–220 (1976)

Thermodynamic Limit of the Free Energy and Correlation Functions of Spin Systems*

Joel L. Lebowitz
Belfer Graduate School of Science
Yeshiva University
New York, N.Y.

Abstract: We give simple proofs of the existence of the thermodynamic limit of the free energy and of equilibrium states for continuous spin systems with "bounded" boundary conditions. For spin-½ Ising systems we show that the infinite volume limit of a state in which there is a field $h_b > 0$ on the boundaries is the same as that obtained from + boundary conditions (independent of the magnitude of h_b). In an appendix with E. Presutti we present stronger results about the existence and uniqueness of equilibrium states for continuous spin systems.

*Based on lectures given at the meeting on "Mathematical Methods of Quantum Field Theory", Marseilles, France, June 1975 and at the conference on "Quantum Dynamics: Models and Mathematics", Bielefeld, Germany, September 1975. Research supported by AFOSR Grant No. 73-2430 and by N.S.F. Grant No. MPS 75-20638

I. Introduction

In this lecture I "continue" my review of the thermodynamic properties and equilibrium states of spin lattice systems. I shall be concerned both with continuous spin variables and with the Ising spin-½ case which I reviewed earlier [1].

The setting will always be the lattice $\mathbb{Z}^\nu$ at each site of which there is a vector 'spin' variable S_i, $i \in \mathbb{Z}^\nu$, $S_i \in \mathbb{R}^n$. Each S_i has associated with it an intrinsic, or free, probability distribution (the same for all S_i) $\rho(S)dS$: $\rho(S)$ depends only on the magnitude of S, $|S|$, and satisfies the bound

$$\int e^{bS^2} \rho(S)dS < \infty, \quad \text{all } b < b_o. \tag{1.1}$$

where, unless otherwise specified, $b_o = \infty$. I shall designate the class of measures satisfying (1.1) by $\mathcal{S}_1$. Typical examples of such distributions are

$\rho(S) = \exp\,[-V_j(S)]$ where V_j is an even polynomial in $|S|$, $j=2n$, $V_j(S) = a_o\, S^{\,j} + a_2\, S^{\,j-2} + \ldots + a_{j-2}\, S^{\,2} + a_j$, $a_o > 0$. For $j \geq 4$, (1.1) is satisfied with $b_o = \infty$, while for $j = 2$, the Gaussian case, $b_o = a_o$.

A more restricted class of distributions, $\mathcal{S}_2 \subset \mathcal{S}_1$, is obtained if $\rho(S)$ is required to have compact support. This may be taken, without loss of generality, to be the unit ball,

$$\rho(S) = 0 \quad \text{if} \quad |S| > 1 \tag{1.2}$$

The simplest case of such a measure is $\rho(S) = K$ (constant) inside the unit ball. A still more restrictive class of measures, $\mathcal{S}_3 \subset \mathcal{S}_2$ is

$$\rho(S) = K\,\delta(|S|-1) \tag{1.3}$$

For n = 1,2,3 these are respectively the Ising spin-½, the rigid rotator and classical Heisenberg models.

The Hamiltonian of the system in a finite domain $\Lambda \subset Z^\nu$ with 'boundary conditions' (b.c.) b_Λ corresponding to specifying the values of the spin variables outside Λ has the form

$$H(\underline{S}_\Lambda;\ b_\Lambda) = -\frac{1}{2} \sum_{i \neq j \in \Lambda} J(i-j)\, S_i\, S_j - \sum_{i \in \Lambda} h_i\, S_i,$$

$$S_\Lambda = \{S_i ; i \in \Lambda\} \tag{1.4}$$

Here J(i-j) is a symmetric n x n matrix, $J(i-j)\, S_i\, S_j = \sum_{\gamma,\delta}^{n} J_{\gamma\delta}(i-j) S_i^\gamma\, S_j^\delta$, and we assume that the interaction has a finite range, $J(\underset{\sim}{x}) = 0$ for $|\underset{\sim}{x}| > R$: h_i is the "magnetic field" at site i, $h_i = h + \sum_{j \notin \Lambda} J(i-j)\, \bar{S}_j$, $h_i\, S_i = \sum_{\gamma=1}^{n} h_i^\gamma\, S_i^\gamma$, and $\{\bar{S}_j\}$ is a specified set of values of S_i for $i \notin \Lambda$. These constitute the set of boundary conditions $\{b_\Lambda\} = b$, i.e. they specify the fields h_i acting on spins near the boundary of Λ. (More generally only the probability distribution of the spins outside Λ need be specified.) The most commonly used b.c. in statistical mechanics are the 'zero' b.c., corresponding to setting $\bar{S}_i = 0$ (this is essentially the Dirichlet b.c. in field theory [2]) and the periodic b.c. We can take the latter into account for Λ a "rectangle", by modifying the definition of $J(\underline{x})$.

The Gibbs probability distribution of spins in Λ at a temperature $T = \beta^{-1}$ is, for a given external field h and b.c. b_Λ,

$$\nu(dS_\Lambda, \beta, h; \Lambda, b_\Lambda) = Z^{-1} \exp\left[-\beta H(S_\Lambda;\ b_\Lambda)\right] \prod_{i \in \Lambda} \left[\rho(S_i) dS_i\right], \tag{1.5}$$

where

$$Z(\beta,h;\Lambda,b_\Lambda) \equiv \exp[|\Lambda|\ \Psi(\beta,h;\Lambda,b_\Lambda)] = \int \exp[-\beta H] \prod_{i\in\Lambda} [(S_i)dS_i] \tag{1.6}$$

Here $|\Lambda|$ equals the number of sites in Λ, and Ψ is (except for a factor $-\beta$) the free energy per site.

Let $\nu(dS_A,\beta,h;\Lambda,b_\Lambda)$ be the projection of the measure (1.5) onto $A\subset\Lambda$, $S_A = \{S_i;\ i\in A\}$. We are interested in the behaviour of the $\nu(dS_A,\beta,h;\Lambda,b_\Lambda)$, or (essentially) equivalently the correlation functions, as $\Lambda\to\mathbb{Z}^\nu$ through some sequence of domains Λ_j. We are also interested in the properties of the free energy per site, $\Psi(\beta,h;\Lambda,b_\Lambda)$ in this limit: here we want to require that the Λ_j be reasonable, e.g. the fraction of sites within a distance R of $\partial\Lambda_j$ $|\partial_R\Lambda_j|/|\Lambda_j|$ go to zero as $j\to\infty$ [3]. We shall denote these limits by $\nu(dS_A,\beta,h;b)$ and $\Psi(\beta,h:b)$. Some of the questions which naturally arise are the existence, uniqueness, analyticity (in β and h), and cluster properties of these limits. Uniqueness refers both to different ways of letting $\Lambda\to\mathbb{Z}^\nu$ for given b.c. and to the dependence of the limit on b.c.

II. Thermodynamic Limit of the Free Energy

Among the above questions the easiest one to tackle is the existence and uniqueness of the limit of $\Psi(\beta,h;\ \Lambda,b_\Lambda)$. It is easy to show, using standard methods, [3] that for $\rho(S) \in \mathcal{S}_2$ (and any reasonable sequence of domain shapes Λ_j), $\Psi(\beta,h;\ \Lambda\ ,b_\Lambda) \to \Psi(\beta,h)$ independent of b.c. The problem is however more difficult in the case where the spins are not bounded. This problem has been treated in some detail by Guerra, Rosen and Simon (GRS) [4] for cases of interest from the point of view of constructive quantum field theory. They were able to prove that various b.c. lead to the same limit for $P(\emptyset)_2$. I shall describe here briefly how to prove such results for the spin lattice systems we are considering here. (I will however not try for the 'best' results which involve considerably more complicated methods: see Appendix for statement of "current" best results). I shall consider first the case of zero b.c., then the case of periodic b.c. and finally the "general" case, which however, as was pointed out by GRS cannot be too general. In all cases the thermodynamic limit will be approached through a sequence of cubes Γ_n, whose sides have lengths $2^n L_o$.

Zero b.c.

Let $Z_n = \exp[|\Gamma_n|\ \Psi_n]$ be the partition function for a cube Γ_n, $|\Gamma_n| = 2^{n\nu}\ L_o^\nu$, with zero b.c. Clearly

$$\Psi_n \leq \log\ [\int \exp[\alpha\ S^2 + hS]\ \rho(S)dS] \qquad (2.1)$$

where $\alpha = \sum_{|\underset{\sim}{x}|\leq R} \alpha(\underset{\sim}{x})$, $\alpha(\underset{\sim}{x}) = \max_{\delta} \sum_{\gamma=1}^{n} \beta|J_{\gamma\delta}(\underset{\sim}{x})|$. The right side of (2.1) is a uniform (in n) bound on Ψ_n whenever $\alpha < b_o$ in (1.1) which we shall assume to be the case. We next wish to show that

$\Psi_{n+1} \geq \Psi_n - \varepsilon_n$ with $\sum \varepsilon_n < \infty$. This will imply the convergence of Ψ_n to a limit $\Psi(\beta,h;b_o)$ [3].

Remark: We note that by Jensen's inequality

$$Z_n = \langle \exp[-\beta H_n] \rangle_I \geq \exp[-\beta \langle H_n \rangle_I] = 1 \tag{2.2}$$

so that $\psi_n \geq 0$. Here the subscript I indicates that the expectations are with respect to the intrinsic measure $\prod [\rho(S_i)dS_i]$. Combining (2.1) and (2.2) shows that $|\Psi_n|$ is bounded and thus we can always get convergence on subsequences. We are after something stronger however.

To obtain the bound $\Psi_{n+1} \geq \Psi_n - \varepsilon_n$ we note that Γ_{n+1} can be thought of as the union of 2^ν cubes $\Gamma_n^{(k)}, k = 1,\dots,2^\nu$. The Hamiltonian of spins in Γ_{n+1} can therefore be divided into a part which contains only interactions between spins in the same cube $\Gamma_n^{(k)}$ and the interaction part between spins in different cubes $\Gamma_n^{(k)}$,

$$H(S_{\Gamma_{n+1}}; b_o) = \sum_{k=1}^{2^\nu} H(S_{\Gamma_n^k}; b_o) + \sum_{k \neq k'=1}^{2^\nu} U_{k,k'}. \tag{2.2'}$$

where

$$U_{k,k'} = -\sum J(i-j)\, S_i\, S_j,\quad i \in \Gamma_n^{(k)},\ j \in \Gamma_n^{(k')} \tag{2.3}$$

Using Jensen's inequality then yields

$$Z_{n+1} \geq (Z_n)^{2^\nu} \exp[-\beta \sum \langle U_{k,k'} \rangle_n] \tag{2.4}$$

where $\langle\rangle_n$ indicates expectations with respect to the product of the Gibbs measures in each $\Gamma_n^{(k)}$. Thus,

$$-\langle U_{k,k'} \rangle_n = \sum J(i-j) \langle S_i \rangle (\beta,h;\Gamma_n,b_o) \langle S_j \rangle (\beta,h:\Gamma_n,b_o),$$

$$i \in \Gamma_n^{(k)},\ j \in \Gamma_n^{(k')} \tag{2.5}$$

Clearly if $\langle U_{k,k'}\rangle_n \leq 0$ then $\Psi_{n+1} \geq \Psi_n$ establishing the desired result. This will be the case if $h = 0$ or if the system is 'properly' ferromagnetic [2]. To deal with the general case it is sufficient to obtain a bound of the form

$$\sum_{i\in\partial_R\Gamma_n} \langle S_i^2\rangle\ (\beta,h;\ \Gamma_n,\ b_o) \leq K_n\ |\partial_R\Gamma_n|$$

such that $\sum K_n\ 2^{-n} < \infty$, since due to the finite range of the interaction we would have

$$\beta\ |U_{k,k'}| \leq \alpha\{\sum_{i\in\partial_R\Gamma_n^{(k)}} S_i^2 + \sum_{j\in\partial_R\Gamma_n^{(k')}} S_j^2\} \tag{2.6}$$

giving $\Psi_{n+1} \geq \Psi_n - \varepsilon_n, \varepsilon_n = \text{Const.}\ (K_n\ 2^{-n})$.

We shall now obtain such a bound (using an idea due to E. Lieb: private communication) for the case where $\rho(S)=\exp[-V_\ell(S)], \ell\geq 4$. (The Gaussian case can of course be treated explicitly). We write for a general domain Λ with b.c. b_Λ,

$$\langle[V_\ell'(S_i)]^2\rangle = -Z^{-1} \int dS_i\ V_\ell'(S_i) \frac{d}{dS_i} \{e^{-V_\ell(S_i)}\}$$

$$\exp[-H(S_\Lambda,b_\Lambda)] \prod_{j\neq i} [\rho(S_j)dS_j]$$

$$= \langle V_\ell'(S_i)\ [\sum_j J(i-j)\ S_j + h_i]\rangle + \langle V_\ell''(S_i)\rangle \tag{2.7}$$

$$\leq \langle V_\ell''(S_i)\rangle + \langle V_\ell'(S_i)h_i\rangle + \alpha\langle[V_\ell'(S_i)]^2\rangle^{1/2}\langle S_m^2\rangle^{1/2}$$

where the prime stands for differentiation, $\langle S_m^2\rangle = \underset{j\in\Lambda}{\text{Max}} \langle S_j^2\rangle$

and we have integrated by parts, used Schwartz's inequality and dropped irrelevant subscripts, etc. Eq. (2.7) remains valid for $i = m$. Using then the fact that for any $k\geq 1$ and $\delta>0$, $\delta|S_i|^k \geq |S_i|^{k-1} + M_\delta$ we obtain

$$\langle S_i^2\rangle_\Lambda \leq K_\Lambda, \quad \forall\ i\in\Lambda \tag{2.8}$$

The dependence on Λ comes in only through the h_i (or $\bar{S}_j$, $j \notin \Lambda$) so that for free boundary conditions (or when the $\bar{S}_i$ are bounded for all $i \in \mathbb{Z}^\nu$) we have a uniform bound $K_\Lambda = K$ and the result is established;

$$\Psi(\beta,h;\Gamma_n, b_o) \xrightarrow[n\to\infty]{} \Psi(\beta,h; b_o), \tag{2.9}$$

for $\rho(S) = \exp[-V_\ell(S)]$. (The method is clearly applicable whenever $\log \rho(S) = \Phi(S) \in C^2$ and $[\Phi'(S)]^2 \to \infty$ as $S \to \infty$ 'rapidly enough' compared to $|\Phi''(S)|$.)

Periodic b.c.

Let us consider as before the sequence of cubes Γ_n but now with periodic b.c. The "translational invariance" of $H(S_{\Gamma_n}; b_P)$ makes this case simpler to treat, Writing

$$H(S_{\Gamma_{n+1}}; b_P) = \sum_{k=1}^{2^\nu} H(S_{\Gamma_n^{(k)}}; b_P) + G \tag{2.10}$$

we have immediately that

$$|G| \le \text{Const.} \sum{}' S_i^2 \tag{2.11}$$

where the prime indicates that the sum is restricted to i near the "boundaries" of the $\Gamma_n^{(k)}$. Thus

$$\Psi(\Gamma_{n+1}, b_P) \ge \Psi(\Gamma_n,b_P) - K_1 \langle S_i^2\rangle(\Gamma_n,b_P)/2^n. \tag{2.12}$$

where we have used the fact that with periodic b.c. $\langle S_i^2\rangle = \frac{1}{|\Lambda|}\sum_{j\in\Lambda} \langle S_j^2\rangle (\Lambda,b_P)$ is the same for all i. But it is quite easy to obtain a uniform bound on the average $\langle S_i^2\rangle$, which is independent of the volume

$$\frac{\gamma}{|\Lambda|} \sum_{i\in\Lambda} \langle S_i^2\rangle (\Lambda,b_P) \le \frac{\gamma}{|\Lambda|} \log \langle \exp \sum S_i^2\rangle$$

$$\leq \log \{\int \exp\,[(\alpha+\gamma)\, S^2 + hS]\, \rho(S)\, dS \Big/ \int \exp[-(\alpha+\gamma)\, S^2 + hS]\, \rho(S)dS\}$$

$$= K_2,$$

where we have used the same estimates as in (2.1). Thus

$$\psi(\beta,h;\, \Gamma_n,\, b_P) \xrightarrow[n \to \infty]{} \psi(\beta,h;b_P).$$

<u>General b.c.</u>
We may write generally

$$H(S_\Lambda;\, b_\Lambda) = H(S_\Lambda;b_o) + U_b$$

where

$$U_b = \sum_{i\in\Lambda}\ \sum_{j\notin\Lambda} J(i-j)\, S_i\bar{S}_j$$

so that

$$|\beta U_b| \leq \alpha\{\sum_{i\in\partial_R\Lambda} S_i^2 + \sum_{j\in\partial_R^+\Lambda} \bar{S}_j^2\}$$

where $\partial_R^+\Lambda\subset\Lambda^c$ is the set of sites in Λ^c within distance R of Λ.

Hence, using the same arguments as before, we obtain that when the $\bar{S}_j$ are <u>bounded</u> then

$$\lim_{n\to\infty} \psi(\beta,h;\Gamma_n,b_{\Gamma_n}) \to \psi(\beta,h;b_o)$$

The same arguments also show that $\psi(\beta,h:b_p) = \psi(\beta,h:b_o)$

III. Equilibrium States

The infinite volume limit of Gibbs states with boundary conditions b will satisfy the Dobrushin, Lanford and Ruelle (DLR) equations,

$$\nu(dS_\Lambda : b) = \int \nu(dS_\Lambda | S_{\Lambda^c})\ \nu(dS_{\Lambda^c} : b), \tag{3.1}$$

where

$$\nu(dS_\Lambda | S_{\Lambda^c}) = \exp[-\beta H(S_\Lambda | S_{\Lambda^c})] \prod_{i \in \Lambda} \rho(S_i)\ dS_i / \text{Normalization} \tag{3.2}$$

is the conditional probability of spins $\underline{S}$ in Λ given $\{S_i; i \in \Lambda^c\}$ and $H(S_\Lambda | S_{\Lambda^c})$ is given by the right side of (1.4) with

$h_i = h + \Sigma_{j \in \Lambda^c} J(i-j) S_j.$

The converse statement; any state satisfying (3.2) may be obtained as the limit of finite volume Gibbs states given by Eq. (1.5) with suitable general boundary conditions (permitting a specified probability distribution for the S_i, $i \in \Lambda^c$), is also true under mild regularity assumption on the solutions of Eq. (3.2) (see Appendix).

The questions regarding the existence and uniqueness of the infinite volume limits of Gibbs states are therefore equivalent to questions regarding the solutions of the DLR equations, e.g. iff (3.2) has a unique (state) solution (for some values of β and h) then $\nu(dS_\Lambda, \beta, h; \Lambda, b_\Lambda)$ will have a limit independent of (reasonable) b.c. Furthermore under the same type of regularity assumptions, the DLR equations are also equivalent to Kirkwood-Salsburg type equations. The latter type equations are particularly useful for proving uniqueness and analyticity of the state at high temperatures and fields [5].

It is clear that for $\rho(S)\in\mathcal{S}_2$ the measures $\nu(dS_A;\Lambda,b_\Lambda)$, having their support in the unit ball of $\mathbb{R}^{n|A|}$, will have subsequences which approach limits as $\Lambda\to\infty$. Similar results will hold for $\rho(S)\in\mathcal{S}_1$ with periodic or more general b.c. such that $|h_i|\leq M,\forall i$, since we then have by (2.8) that $\langle S_j^2\rangle_{\Lambda,b_\Lambda} \leq K$ uniformly in Λ, and this implies that for any $\varepsilon > 0$ we can find a ball $B_\varepsilon\subset\mathbb{R}^{n|A|}$ such that $S_A = \{S_i: i\in A\}\in B_\varepsilon$ with probability $\geq (1-\varepsilon)$.

The requirement that h_i, i.e. $S_j, j\notin\Lambda$, be bounded is however too strong a restriction since it will clearly not be satisfied by the "natural" b.c.occurring in (3.1); see Appendix.

Uniqueness-Analyticity-Decay of Correlations

It has been shown by Israel [5] that for $\rho(S)\in\mathcal{S}_2$ there is a unique equilibrium state at sufficiently high temperatures and large magnetic fields. This state, which being unique is of necessity translation invariant, has correlation functions analytic in β and h, which (for finite range potentials) decay exponentially. Similar results were proven a long time ago for Ising spins and for continuum systems [3] where "large magnetic field" corresponds to "small fugacity". We expect similar results to hold also for $\rho(S)\in\mathcal{S}_1$: indeed they have been proven for some cases of interest in field theory [6].

Since these results do not hold at phase transitions we cannot expect to extend them beyond such a limited region in the β-h plane without further restrictions on the interactions $J(\underline{x})$ and the $\rho(S)$. A particularly interesting class of systems, of importance both in statistical mechanics and in field theory, are ferromagnetic spin systems for which such nice results ought to (and in many cases have been shown to) hold for all h = 0. This can indeed be shown to be the case; see Appendix.

IV. Ferromagnetic Ising Spins with Boundary Fields

From the point of view of the DLR equations we need only consider b.c. which are superpositions of 'pure' b.c., b_Λ, corresponding to a specification of S_i, $i \in \Lambda^c$, with $S_i \in$ Supp. $\rho(S)$. For some purposes however it is useful to consider also 'unrestricted' b.c. where we simply specify the h_i's, $i \in \partial_R \Lambda$, without regard as to whether they actually come from properly specified spins outside Λ. Thus zero b.c. for $\rho(S) \in \mathcal{S}_3$ are of this type. Periodic b.c. as well as those of interest in field theory may also be put in this category of b.c. In any case however a state obtained with such b.c. should again satisfy the DLR equations.

We shall now show that for a spin-½ ferromagnetic Ising system $J(x) \geq 0$, the infinite volume equilibrium state obtained by putting on an external field $h_b > 0$ on the boundary spins is identical to ν_+; the state obtained from + boundary conditions. To emphasize the fact that we are now dealing with the simple Ising system we shall use $\sigma_i = \pm 1$, instead of S_i, for our variables. Let us denote by $\langle \sigma_A \rangle (h_b;\Lambda)$ the expectation value of $\sigma_A = \prod_{i \in A} \sigma_i$ in a region $A \subset \Lambda$ in the presence of a boundary field h_b. As we already know, when there is an external uniform field $h \neq 0$ or when the temperature is above the critical temperature for spontaneous magnetization there is a unique equilibrium state [1] and hence we need only be concerned with the case $h = 0$, and $\langle \sigma_i \rangle_+ = m > 0$, where $<\sigma_1>_+ = \lim_{h \downarrow 0} <\sigma_1> (h) = \lim_{\Lambda \to \infty} <\sigma_1> (h_+;\Lambda)$.

We first note that + boundary conditions correspond to a field h_i^+ for $i \in \partial_R \Lambda$, the boundary of Λ with $h_i^+ < \alpha = h^+$. Thus by the GKS inqualities [1,2,3] $\langle \sigma_A \rangle (h_b;\Lambda) \geq \langle \sigma_A \rangle_+ (\Lambda)$ if $h_b \geq h^+$. On the other hand letting $h_b \to \infty$ clearly corresponds to putting + b.c.

on the region $\Lambda^- = \Lambda \setminus \partial_R \Lambda$. We therefore have, again by G.K.S., that for $A \subset \Lambda^-$,

$$\langle\sigma_A\rangle_+ (\Lambda^-) \geq \langle\sigma_A\rangle (h_b;\Lambda) \geq \langle\sigma_A\rangle_+ (\Lambda) \text{ for } h_b > h^+ \quad (4.1)$$

Letting $\Lambda \to Z^\nu$ then yields [7]

$$\langle\sigma_A\rangle (h_b) \equiv \lim_{\Lambda\to\infty} \langle\sigma_A\rangle (h_b;\Lambda) = \langle\sigma_A\rangle_+ \text{ for } h_b > h^+ \quad (4.2)$$

Our strategy for proving (4.2) for all $h_b > 0$ will be to show that $\langle\sigma_A\rangle (h_b)$ is analytic in h_b for Re $h_b > 0$. It will actually be sufficient to prove the result just for $\langle\sigma_i\rangle (h_b)$ since by the FKG inequalities [7],

$$0 \leq \langle\rho_A\rangle_+ (\Lambda^-) - \langle\rho_A\rangle (h_b;\Lambda) \leq \sum_{i\in A} [\langle\rho_i\rangle_+ (\Lambda^-) - \langle\rho_i\rangle (h_b;\Lambda)] \quad (4.3)$$

where $\rho_i = \frac{1}{2} (1+\sigma_i)$, $\rho_A = \prod_{i\in A} \rho_i$.

We remark that (i) by GHS $\langle\sigma_i\rangle (h_b)$ is a concave function of h_b so that by (4.2) $\langle\sigma_i\rangle (h_b) > 0$. (ii) If σ_A is replaced by ρ_A then (4.1) and (4.2) remain valid in the presence of non-uniform fields.

The argument so far is valid for all measures in $\mathcal{J}_2$. To obtain analyticity in h_b we have to restrict ourselves to the Ising spin ½ case. We use the Lee-Yang theorem in a manner similar to its use by Lebowitz and Penrose [8] for proving analyticity in a uniform external field. Letting $\eta = \exp [2\beta h_b]$, $\xi_i = \exp [2\beta h_i]$ (assuming for the moment that there is an external field present at site i) the grand partition function in Λ can be written (up to factors which are irrelevant) in the form

$$\Xi(\eta, \xi_i;\Lambda) = P(\eta,\Lambda) + \xi_i Q(\eta;\Lambda) \quad (4.4)$$

where P and Q are polynomials in η.

We then obtain [8], when $\xi_1 = 1$,

$$\langle\sigma_i\rangle (h_b;\Lambda) = [1 - \Phi(\eta;\Lambda)]/[1 + \Phi(\eta,\Lambda)]$$

where $\Phi(\eta;\Lambda) = P(\eta,\Lambda)/Q(\eta;\Lambda)$. By the Lee-Yang theorem $\Xi(\eta,\xi_i;\Lambda) \neq 0$ if $|\xi_i| > 1$ and $|\eta| > 1$. Hence $|\Phi(\eta;\Lambda)| \leq 1$ when $|\eta| > 1$: $\Phi(\eta;\Lambda)$ is thus a bounded analytic function outside the unit disk and

$$\Phi(\eta;\Lambda) = [1 - \langle\sigma_i\rangle (h_b;\Lambda)]/[1+\langle\sigma_i\rangle (h_b;\Lambda)] \xrightarrow[\Lambda\to\infty]{} \frac{1-m}{1+m},\ h_b \geq h^+$$

Therefore by Vitali's theorem $\lim_{\Lambda\to\infty} \Phi(\eta;\Lambda)$ is constant for $|\eta| > 1$ which proves the result.

References

1. J.L. LEBOWITZ, in Mathematical Problems in Theoretical Physics H. ARAKI, Ed. (Springer-Verlag, 1975)
2. B. SIMON, The $P(\Phi)_2$ Euclidian Quantum Field Theory, (Princeton University Press, 1975); Constructive Quantum Field Theory G. VELO and A. WIGHTMAN, Eds. (Springer-Verlag, 1974).
3. D. RUELLE, Statistical Mechanics, (Benjamin, 1969)
4. F. GUERRA, L. ROSEN and B. SIMON, Ann. of Math. 101, 111 (1957) Boundary Conditions for $P(\Phi)_2$, Preprint.
5. R.B. ISRAEL, High Temperature Analyticity in Classical Lattice Systems, Preprint.
6. See articles by J. GLIMM, A. JAFFE and T. SPENCER in ref. 2b.
7. J.L. LEBOWITZ and A. MARTIN-LÖF. Comm. Math. Phys. 25, 276 (1972).
8. J.L. LEBOWITZ and O. PENROSE, Comm. Math. Phys. 11, 99 (1968)

Note: The following note was submitted to The Fourth International Symposium on Information Theory, which will be held in Leningrad in June 1976. The notation is slightly different from that used in the body of the paper. In particular $\rho(dS) = \exp[\Phi(S)]\ \mu(dS)$ where $-\Phi$ is the self-interaction.

Appendix:

STATISTICAL MECHANICS OF CONTINUOUS SPIN SYSTEMS

by

J.L. LEBOWITZ AND E. PRESUTTI*

Abstract:
We present results relating to the existence and uniqueness of the free energy equilibrium states for classical continuous spin systems with superstable interactions.

We consider the lattice $\mathbb{Z}^\nu$ at each site of which there is a vector spin variable S_x, $x \in \mathbb{Z}^\nu$, $S_x \in \mathbb{R}^d$. We denote by $\underline{S} \in \{S\}$ a spin configuration on $\mathbb{Z}^\nu$. Each S_x has associated with it an intrinsic positive measure $\mu(dS)$, the same for all sites, such that $\int \mu(dS)\, e^{-\alpha S^2} < \infty$ for $\alpha > 0$. The energy of a given spin configuration S_Λ in $\Lambda \subset \mathbb{Z}^\nu$ consists of both pair and self interactions and satisfies the following conditions:

a) Superstability There exists $A > 0$, $C \in R$ such that

$$U(S_\Lambda) \geq \sum_{x \in \Lambda} [A\, S_x^2 - C] \tag{1}$$

where S_Λ is a configuration in Λ.

* permanent address: Istituto Matematico
Universita Dell'Aquila, L'Aquila, Italy

b) Regularity If Λ_1, Λ_2 are disjoint then their interaction energy $W(S_{\Lambda_1}|S_{\Lambda_2}) = U(S_{\Lambda_1} \otimes S_{\Lambda_2}) - U(S_{\Lambda_1}) - U(S_{\Lambda_2})$ has the bound

$$|W(S_{\Lambda_1}|S_{\Lambda_2})| \leq \frac{1}{2} K \sum_{x \in \Lambda_1} \sum_{y \in \Lambda_2} |S_x| \, |S_y| \, |x-y|^{-\nu-\varepsilon} \qquad (2)$$

where $|x| = \max_{1 \leq i \leq \nu} |x^i|$, $|S| = [\sum_{i=1}^{d} (S^i)^2]^{1/2}$

For Λ bounded in $\mathbb{Z}^\nu$ we consider the restriction, S_{Λ^c} of $\underline{S}$ to Λ^c and define the partition function $Z(\Lambda|S_{\Lambda^c})$ and free energy per site $F(\Lambda|S_{\Lambda^c})$ with 'boundary conditions' determined by $\underline{S}$ as

$$Z(\Lambda|S_{\Lambda^c}) = \int \mu_\Lambda(dS_\Lambda) \exp [-U(S_\Lambda) - W(S_\Lambda|S_{\Lambda^c})]$$

$$F(\Lambda|S_{\Lambda^c}) = |\Lambda|^{-1} \ln Z(\Lambda|S_{\Lambda^c})$$

where $\mu_\Lambda(dS_\Lambda) = \prod_{x \in \Lambda} \mu(dS_x)$, $|\Lambda|$ = # of sites in Λ. (The dependence on temperature and magnetic field is included in U and W; it will be made explicit when necessary.)

Theorem 1. Let (1) and (2) hold and let $\underline{S} \in \mathcal{H}_a$:

$\mathcal{H}_a = \{\underline{S} | S_y^2 \leq a \ln |y| \text{ for } |y| > 1\}$. Let $\{\Lambda\}$ be a sequence of increasing domains tending to $\mathbb{Z}^\nu$ in the sense of Van Hove [1] then $\lim_{n\to\infty} F(\Lambda_n|S_{\Lambda^c}) = F$ exists and is independent of the sequence $\{\Lambda_n\}$ and of the b.c. S_{Λ^c}.

Remark: "Zero" b.c. correspond to $S_x = 0$ for $x \in \Lambda^c$. The thermodynamic limit of the "periodic" b.c. free energy can also be shown to exist and be equal to F.

A probability measure ν on the configuration space $\{S\}$ is said to be regular if it satisfies the following condition: There exists $\gamma>0$, $\delta\geq 0$, such that for every Δ bounded in Z^{ν} and $N^2>0$ the following holds:

$$\nu[B(N^2|\Delta)] \leq \exp\,[-|\Delta|\ (\gamma\ N^2 - \delta]$$

where

$$B(N^2|\Delta) = \{\underline{S} \mid \sum_{x\in\Delta} S_x^2 \geq N^2\ |\Delta|\}$$

For Λ bounded in $\mathbb{Z}^{\nu}$ we denote by $F(\Lambda|\nu)$ the free energy in Λ for boundary conditions specified by the measure ν as

$$F(\Lambda|\nu) = \int \nu(d\underline{S})\ F(\Lambda|S_{\Lambda^c})$$

<u>Theorem 2.</u>
Let (1) and (2) hold and let Λ_n be as in theorem 1 then for ν regular $\lim_{n\to\infty} F(\Lambda_n|\nu) = F$.

The finite volume equilibrium measure with boundary conditions S_{Λ^c}, $\nu_\Lambda(dS_\Lambda|S_{\Lambda^c})$ is given by

$$\nu_\Lambda(dS_\Lambda|S_{\Lambda^c}) = Z^{-1}\ (\Lambda|S_{\Lambda^c})\ \mu(dS_\Lambda)\ \exp[-U(S_\Lambda)\ -W(S_\Lambda|S_{\Lambda^c})] \tag{3}$$

A measure ν on $\{S\}$ is said to be an equilibrium measure (for our system) if its conditional probabilities $\nu(dS_\Lambda|S_{\Lambda^c})$ satisfy the Dobrushin, Lanford and Ruelle (DLR) [2] equations, i.e. eq. (3).

<u>Theorem 3.</u>
Let the conditions of Theorem 1 be satisfied and let $\nu_{\Lambda_n}(dS_{\Lambda_n}|S_{\Lambda_n^c})$ be finite volume equilibrium states: it is always possible to choose subsequences n_i (which may depend on the b.c.) such that $\nu_{\Lambda_{n_i}}(dS_{\Lambda_{n_i}}|S_{\Lambda^c_{n_i}}) \to \nu$ a <u>regular</u> equilibrium measure on $\{S\}$, [3].

The one component spin system, $S_x \in R$, will be called ferromagnetic with translation invariant interaction if

$$U(S_\Lambda) = -\frac{1}{2} \sum_{x \neq y \in \Lambda} J(x-y)\, S_x S_y - \sum_{x \in \Lambda} \Phi(S_x) -$$

$$- h \sum_{x \in \Lambda} S_x ;\; J(x) \geq 0,$$

Theorem 4.

Let ν be a regular equilibrium measure of a ferromagnetic system in an external field h whose interactions satisfy (1) and (2) then ν is unique (and hence translation invariant) whenever the infinite volume free energy F(h) is differentiable with respect to h [4].

Acknowledgments

Work supported by NSF Grant # MPS 75-20638. We would like to thank Prof. David Ruelle for very valuable discussions.

References

1. D. RUELLE, Statistical Mechanics (Benjamin, 1969).
2. L. DOBRUSHIN, Funct. Appl. 2, 291 (1968); O.E. LANFORD and R. RUELLE, Comm. Math. Phys. 13, 194 (1969).
3. D. RUELLE, Probability Estimates for Continuous Spin Systems (Preprint): The bounds obtained in this paper form the basis for our results..
4. J.L. LEBOWITZ and A. MARTIN-LÖF, Comm. Math. Phys., 25, 276 (1972) prove this theorem for spin -$\frac{1}{2}$ Ising systems.

Acta Physica Austriaca, Suppl. XVI, 221–239 (1976)

The Lorenz Attractor and the Problem of Turbulence†

David Ruelle*

Institute for Advanced Study, Princeton, N.J.

*Permanent address: Institut des Hautes Etudes Scientifiques, 91440 Bures-sur-Yvette, France.

†Part of this work was performed while the author was visiting Princeton University and Yeshiva University.

I. Introduction

Turbulence in the flow of liquids is a fascinating phenomenon. This may partly explain the conceptual confusion which exists in the scientific literature as to the nature of this phenomenon. My own opinion, and that of some other people, is that turbulence at low Reynolds numbers corresponds to a mathematical phenomenon observed in the study of solutions of differential equations

$$dx/dt = X(x) \tag{1}$$

The equation just written has to be understood as a time evolution equation in several dimensions. The mathematical phenomenon referred to is that in many cases, solutions of (1) have an asymptotic behavior when $t \to \infty$ which appears erratic, chaotic, "turbulent", and the solutions depend in a sensitive manner on initial condition.

According to conventional ideas, when the Reynolds number of a fluid is increased a number of independent frequencies, or "oscillations", successively appear in the fluid. The superposition of a sufficiently large number of frequencies would produce turbulence. This view was developed by E. Hopf [4], who explained the occurrence of various frequencies by successive "bifurcations". A more popular version of the same ideas is due to Landau [6]. By a "superposition of k frequencies" is here meant a quasi periodic time dependence of the form

$$x(t) = F(\omega_1 t, \ldots, \omega_k t) \tag{2}$$

where F is periodic of period 2π in each argument separately, and the ω_i are k irrationally related frequencies. Such a quasi periodic motion does appear erratic, chaotic, or turbulent, but does not depend in a sensitive manner on initial condition. F. Takens and myself [11] noted however that for $k \geq 4$, a small perturbation could transform a quasi periodic motion into one that would again look "turbulent", and would furthermore have a sensitive dependence on initial conditions. This can be interpreted in physical terms by saying that putting a nonlinear coupling between 4 or more oscillations can produce a "turbulent" time evolution with sensitive dependence on initial condition. Technically we proved that "strange Axioms A attractors" could occur. Axiom A flows and attractors have been introduced by Smale, and are very nice mathematical objects to work with from an abstract point of view. Instead of discussing those I shall however turn now to the discussion of an object which has greater geometric simplicity, and intuitive appeal: The Lorenz attractor.

Lorenz' work [8], which was unfortunately overlooked in [11], is the first attempt at interpreting turbulence by solutions of differential equations which appear chaotic, and have sensitive dependence on initial condition. The ideas of Lorenz and those of Takens and myself have recently received support from the theoretical work of McLaughlin and Martin [9] and the experimental work of Gollub and Swinney [3]. It can be hoped that more experimental results on the onset of turbulence in various systems will become available in the next few years; their theoretical interpretation will constitute a worthy challenge for the mathematical physicist.

II. The Lorenz Attractor

2.1. Generalities. The Lorenz equations are the following

$$\begin{aligned} \dot{x} &= -\sigma x + \sigma y \\ \dot{y} &= -xz + rx - y \\ \dot{z} &= xy - bz \end{aligned} \qquad (3)$$

where $\dot{x} = dx/dt$, etc. and σ, r, b are positive numbers. These equations are obtained as an approximation to partial differential equations describing convexion in a fluid layer heated below (Bénard problem). The unknown functions in the Bénard problem are expanded in Fourier series and an infinite system of coupled differential equations is obtained. By putting all Fourier coefficients equal to zero except three, one gets the truncated system (3). The system (3) is of course a badly mutilated version of the original partial differential equations, and one may well wonder what relation its solutions have to the original problem. Things may not be as bad as they appear, but we won't go into that question. We shall interest ourselves here only in mathematical phenomena exhibited by solutions of (3): their chaotic appearance and sensitive dependence on initial condition. These phenomena occur for a suitable range of values of the parameters σ, r, b and Lorenz made for his numerical studies the choice

$$\sigma = 10, \quad b = 8/3, \quad r = 28 \qquad (4)$$

2.2. There is a bounded region B of $\mathbb{R}^3$ such that every solution of (3) eventually becomes trapped in B.
Lorenz notices that this holds for a general system of the form

$$dx_i/dt = \sum_{j,k} a_{ijk}\, x_j x_k - \sum_j b_{ij} x_j + c_i$$

where $\sum a_{ijk}\, x_i x_j x_k$ vanishes identically and $\sum b_{ij} x_i x_j$ is positive definite. One verifies indeed immediately that

$$\frac{1}{2}\frac{d}{dt} \sum_i x_i^2 = - \sum_{ij} b_{ij} x_i x_j + \text{lower order}$$

is negative when $\sum_i x_i^2$ is large enough. The system (3) satisfies the above assumptions after the change of variables

$$x' = x,\ y' = y,\ z' = z - r - \sigma$$

2.3. <u>The time evolution given by (3) contracts volumes in $\mathbb{R}^3$ at a constant rate.</u>

$$\frac{\partial \dot{x}}{\partial x} + \frac{\partial \dot{y}}{\partial y} + \frac{\partial \dot{z}}{\partial z} = -(\sigma + b + 1)$$

Notice that for the values (4) of the parameters this is $-13\frac{2}{3}$, which is a very fast rate.

2.4. <u>The system (3) is invariant under the symmetry</u>

$$(x,y,z) \to (-x, -y, z)$$

2.5. <u>Steady state solutions and bifurcations</u>
Clearly, the right-hand side of (3) vanishes at the point $O = (o,o,o)$, which is thus a steady state solution. For $r < 1$ this is the only steady state solution and it is attracting. This is seen by looking at the matrix of partial derivatives of the right-hand side of (3) at the point O:

$$\left(\frac{\partial x_i}{\partial x_j}\right) = \begin{pmatrix} -\sigma & \sigma & \\ r & -1 & \\ & & -b \end{pmatrix}$$

The eigenvalues of this matrix have negative real part (are in fact real negative), showing that O is attracting.

When r becomes larger than 1, O loses its attracting character (one eigenvalue becomes positive) and two new steady state solutions appear:

$$C = (\sqrt{b(r-1)},\ \sqrt{b(r-1)},\ r-1)$$

$$C' = (-\sqrt{b(r-1)},\ -\sqrt{b(r-1)},\ r-1)$$

O, C, C' are the only steady state solutions for r > 1. The bifurcation which leads to the creation of C, C' is illustrated in Fig. 1. This picture shows that C, C' should be attracting. Looking at the matrix of partial derivatives $(\partial x_i/\partial x_j)$ at C or C' confirms this: the eigenvalues have a negative real part: first they are real negative, then one pair becomes complex conjugate. However if

$$r > \sigma(\sigma+b+3)(\sigma-b-1)^{-1} > 0 \qquad (5)$$

then the complex conjugate pair has now a positive real part. The values (4) satisfy (5). We have thus a situation where there are three steady state solutions, none of them attracting. Let us examine the situation in detail.

(a) near O points come in along a two-dimensional surface (stable manifold of O) and go out along a line (unstable manifold)

(b) near C points come in along a line and go out along a two-dimensional surface: they spiral out because the eigenvalues with positive real part have a non vanishing imaginary part

(c) similarly near C'.

In this situation, numerical calculations by Lorenz showed that the solutions of (3) an apparently erratic, chaotic, or turbulent behavior.

2.6. <u>The Poincaré map</u>

To understand what is happening, it is convenient to go from 3 to 2 dimensions, by using a Poincaré map. Let P be a point in the plane z = 27 containing C and C'. We assume that dz/dt > 0 at P, i.e. P is inside an equilateral hyperbola through C and C'.

We follow the integral curve through P, and let ϕP be the next point at which this curve crosses the plane $z = 27$, going downwards. The map $\phi: P \to \phi P$ is our Poincaré map. Oscar Lanford put the problem to a computer, and Fig. 2 is the answer. It is seen that the successive images of a point P by the Poincaré map ϕ tend to lie on two arcs Γ and Γ' *). The point C, C' are respectively on the continuation of Γ and Γ'. Also drawn is the line Σ of points which do not come back to cross the plane $z = 27$, because the integral curve they determine goes to O (Σ is part of the unstable manifold of O).

Because of the symmetry 2.4., it suffices to understand the action of ϕ on Γ. Since points close to C are spiraling away from C, it is understandable that the piece of Γ contained between C and Σ is stretched in the manner described by Fig. 3. The computer tells us that the piece of Γ beyond Σ is mapped onto Γ' as indicated in Fig. 4. Notice that, using the notation of Fig. 3, 4, ϕB is not defined, and if $x \to \Sigma$ then ϕx tends to B' or B" depending on whether x is on the same side of Σ as C, or on the other side **).

2.7. Γ and Γ' are not line arcs

I must apologize for having implied that Γ and Γ' are line arcs. A little thinking shows that this would be in contradiction with the uniqueness of solutions of differential equations. To see better what happens, it is convenient to use the symmetry 2.4. and identify symmetric points. The map ϕ is now well defined at B since B' and B" are identified. Figs. 3,4, become Fig. 5. This describes the image under ϕ of a line arc approximating Γ. It so happens that the two line arcs on the picture of the right are practically on top of each other. This is not too astonishing in view of 2.3. Altogether one expects that when $n \to \infty$, the points $\phi^n P$ tend to sets Γ and Γ' which look like Cantor sets in cross section.

*) The point P itself might lie anywhere. It has therefore not been drawn, neither have the first few points $\phi^k P$.

**) The points B', B" are on the unstable manifold of O, cf section 2.5(a).

2.8. Sensitive dependence on initial condition

It should now be clear that the map ϕ compresses Γ Γ' in one direction and stretches it in another. Therefore given two points P and P' close together, in general the stretching will have the effect that $\phi^n P$ and $\phi^n P'$ will be farther apart. In fact, their distance will increase exponentially with n, as long at least as it is not too large. Better than that (or worse than that) after a while $\phi^n P$ and $\phi^n P'$ will fall on different sides of Σ, and from then on their futures are totally dissimilar.

Going back to the system (3) we see that its solutions depend sensitively on initial condition. This would not be true for a quasi-periodic motion (2) as one readily checks.

2.9. Hyperbolicity

The combination of compressing in one direction and stretching in another is called hyperbolicity. We refrain from a formal definition. A little thinking shows that hyperbolicity causes sensitive dependence on initial condition. This is exactly what happens for Axiom A time evolutions, which are defined by a hyperbolicity requirement. Technically however, the Lorenz equations do not satisfy Axiom A. That is because the integral curves can come arbitrarily close to O, and are then "slowed down" for an arbitrarily long time: this spoils some uniformity in the hyperbolicity required for Axiom A to hold.

2.10. The work of Lanford and Guckenheimer

The facts mentioned above about the solutions of (3) have been obtained numerically by use of a computer by Lorenz and by Lanford. Unfortunately (fortunately for mathematicians) computers do not yet prove theorems. Therefore I must make the reservation that, while one is quite confident that things are as described above, there are no proofs yet, and proofs may be hard to obtain. Work in this direction is being done by Lanford*) and Guckenheimer**) Fig. 5 suggests the study of maps of a line segment onto itself

---

*) O. Lanford, private communication

**) J. Guckenheimer, A Strange, Strange Attractor. Preprint

of the type Fig. 6. Such maps have been investigated by Li and Yorke [7], and by W. Thurston (unpublished). From this work one can derive information on periodic orbits of (3).

2.11. Hysteresis

Notice that the sets Γ and Γ' in Fig. 2 do not extend all the way to C and C' respectively. This is understandable since ϕ actually pushes Γ away from C, and the piece of Γ near C has to be re-fed every time from Γ'. When the parameter r is decreased it is thus possible that $\Gamma \quad \Gamma'$ will remain an attracting set for ϕ, but that C and C' also become attracting. Depending on the initial condition the system will thus be in $\Gamma \quad \Gamma'$, or in C, or in C'. A corresponding situation will prevail for the system (3) for a certain range of r. In fact the system will be near a different attracting set depending on whether r is raised from low values or decreased from high values. This phenomenon is called <u>hysteresis</u>. In the case at hand, I do not expect hysteresis to be important, due to the smallness of the gap between Γ and C.

III. Questions

An attractor for a flow (or differential equation) is a compact set Λ such that all points sufficiently close to Λ tend to Λ under time evolution when the time tends to $+\infty$. To this definition some "irreducibility" condition should be added; for Axiom A flows it is required that Λ be connected. The following striking result holds.

3.1. Theorem. Let $t \to x(t)$ be the time-evolution for a C^2 Axiom A flow. Then for almost all initial condition $x(o)$ with respect to Lebesgue measure*), $x(t)$ tends to an attractor**) when $t \to +\infty$. The open set of those $x(o)$ such that $x(t)$ tends to a given attractor Λ is called the basin of Λ.
There is a probability measure μ with support Λ such that for almost all $x(o)$ with respect to Lebesgue measure in the basin of Λ, and every continuous function ϕ on this basin

$$\lim_{T \to \infty} \frac{1}{T} \int_0^T \phi(x(t))dt = \int \mu(dy)\phi(y) \tag{6}$$

The measure μ is invariant and ergodic under time evolution.

It is a natural question whether this carries over to the Lorenz attractor. My guess is that it does. Notice that (6) looks like the ergodic theorem but the formula holds for almost all $x(o)$

*) We call Lebesgue measure the measure defined by a Riemann metric, it is thus not unique, but sets of Lebesgue measure zero are unambiguously defined.

**) An Axiom A flow has a finite number of attractors

with respect to Lebesgue measure, and the measure μ in the right-hand side is not Lebesgue measure. Theorem 3.1. is proved in [1], which also contains a characterization of μ by a variational principle. Would this variational principle apply to the Lorenz attractor?*)

If (6) holds one can define a time correlation function

$$F(t) = \int \mu(dx(o))\phi(x(o))\phi(x(t)) - \left(\int \mu(dx)\phi(x)\right)^2$$

where ϕ is assumed to be differentiable.

3.2. Conjecture. For the Lorenz attractor, $F(t)$ tends to zero exponentially fast when $t \to \infty$.

The corresponding fact is known to be true in the discrete-time Axiom A case (Axiom A diffeomorphisms [10]). It expresses the sensitive dependence on initial condition (or the loss of information on initial condition). In any case one expects that $F(t)$ tends to zero, even if not exponentially (mixing).

---

*) If it does, this permits an estimate of the entropy of the flow.

IV. Some Misconceptions

There is now detailed experimental evidence on the onset of turbulence. Gollub and Swinney [3] have indeed studied an example of transition to turbulence with the following features:

(a) The transition is sharp and without hysteresis (within experimental error; there is probably some hysteresis).

(b) The nature of the correlation functions changes abruptly*).

The oscillations predicted by the quasi-periodic Landau picture are not present.

These results and the analysis of McLaughlin and Martin [9] indicate that the ideas of Lorenz [8], Takens and myself [11] are in better agreement with experiment than traditional interpretations of turbulence.

In view of all this I want to discuss critically some ideas, which appear frequently in the literature, and which I feel to be misconceptions.

4.1. <u>That an external source of "random noise" is necessary to explain the apparent loss of information in turbulent flow.</u>
The sensitive dependence on initial condition seems to explain all what has to be explained. Of course there is some noise in experiments. There are now theorems to the effect that for Axiom A systems a little noise only changes a little bit the measure μ of (6) [see Sinai [12], Kifer [5]]. Whether the effect of the noise will be weak or important will depend on whether the at-

*) The frequency spectrum shows an abrupt transition from a regime with sharp spikes, to one with only continuous spectrum.

tractor one is on is strongly or weakly attracting*).

4.2. <u>That a measure describing turbulence should be approximately Gaussian</u>.

This idea seems to stem from the notion that in turbulent motion a large number of different frequencies of small amplitude are superposed incoherently. This comes from the Landau picture of turbulence. A moment's thinking shows however that according to the Landau picture different frequencies are in different directions of the "phase space" of the system (functional space of velocity fields), and don't therefore superpose incoherently. On the other hand it is clear that the Lorenz attractor does not carry an approximately Gaussian measure (cf Fig. 2). Also the asymptotic measure μ on an Axiom A attractor defined by (6) cannot be approximately Gaussian because one can prove [1] that Axiom A attractors have Lebesgue measure zero.

4.3. <u>That the condition of stationarity is more or less sufficient to determine the measure describing turbulence</u>.

Let μ be a measure describing the statistical, or time-averaged, behavior of a turbulent flow (cf equation (6)) - then μ is of course invariant under time evolution. However, on a strange**) Axiom A attractor there is a continuum of different probability measures invariant under time evolution. To fish the right one out one needs some extra information which can, in the Axiom A case, be given by a variational principle (cf [1]).

To summarize, I think it would be a miracle if the usual procedure of imposing stationarity, truncating the resulting system of equations, and looking for a Gaussian solution, would lead to results much related to physics.

---

*) Near a normal bifurcation the attracting character of an attractor is weak, and therefore "noise amplification" should occur.

**) We call an Axiom A attractor "strange" if it does not reduce to a fixed point or a closed orbit of the flow.

V. Outlook

There remains a lot to be done to have a good understanding of the onset of turbulence. Let me mention two problems which obviously deserve investigation.

5.1. Study of asymptotic measures on general attractors

The conjecture is that for almost all initial points with respect to Lebesgue measure the time average (6) makes sense (at least for almost all time evolutions-in some sense). The problem would be to study the measures μ thus obtained. How generally do the time correlation functions decrease exponentially?

Notice that the existence of (6) for almost all initial points is proved for Axioms A flows [1], flows asymptotic to a torus with a quasi-periodic motion on it (because of unique ergodicity), and Hamiltonian flows on compact manifolds (this follows from the invariance of the Liouville - Lebesgue measure and the ergodic theorem -- I am indebted to D.S. Ornstein for this remark).

5.2. Study of bifurcations occurring in concrete problems at the onset of turbulence

This is a difficult but exciting problem. The only existing work in this direction is that of McLaughlin and Martin [9], which is excellent although perhaps a bit over enthusiastic. As new experimental work becomes available (like the beautiful recent work of Gollub and Swinney [3]), we realize our lack of information on obvious theoretical questions. For instance, in spite of the work of Lorenz and Lanford, simple questions about the Lorenz attractor are still unanswered: how much hysteresis is there? what is the frequency spectrum like? Also what are simple attractors present in low dimension apart from the Lorenz attractor?

In conclusion let me express my feeling that, after decades of misconceptions, we are beginning to have correct ideas on the time dependence in turbulence near its onset. This solves the problem of turbulence only in part, however. In particular the space dependence has striking features (see [2]) which will probably require new conceptual ideas for their elucidation.

Acknowledgments

This work was performed during the summer of 1975, while the author was visiting the Institute for Advanced Study, Princeton University, and Yeshiva University. Thanks are due to S. Adler, A. S. Wightman, and J. L. Lebowitz for these invitations. I am also very indebted to O. Lanford whose work on the Lorenz attractor I have freely used in the present report.

References

1 R. BOWEN and D. RUELLE. The ergodic theory of Axiom A flows. Inventiones math. To appear.

2 R. P. FEYNMAN, R. B. LEIGHTON, and M. SANDS. The Feynman lectures on physics. 2. Addison-Wesley, Reading, Mass., 1964.

3 J. P. GOLLUB and H. SWINNEY. Onset of turbulence in a rotating fluid. Preprint.

4 E. HOPF. A mathematical example displaying features of turbulence. Commun. pure appl. Math. 1 , 303 - 322 (1948).

5 Ju.I. KIFER. On small random perturbations of some smooth dynamical systems. Izv. Akad.Nauk SSSR.Ser. mat. 38 N°5, 1091-1115 (1974). English translation: Math. USSR Izvestija 8, 1083-11o7 (1974)

6 L. D. LANDAU and E. M. LIFSHITZ. Fluid Mechanics. Pergamon, Oxford, 1959.

7 T.-Y. LI and J. YORKE. Period three implies chaos. Preprint

8 E. N. LORENZ. Deterministic nonperiodic flow. J. atmos. Sci. 20, 130 - 141 (1963).

9 J. B. MCLAUGHLIN and P. C. MARTIN. Transition to turbulence of a statically stressed fluid system. Phys. Rev. Lett. 33, 1189 - 1192 (1974), Phys. Rev. A 12, 186 - 203 (1975).

10 D. RUELLE. A measure associated with Axiom A attractors. Amer. J. Math. To appear.

11 D. RUELLE and F. TAKENS. On the nature of turbulence. Commun. Math. Phys., 20, 167 - 192 (1971), 23, 343 - 344 (1971).

12 Ia. G. SINAI. Gibbs measures in ergodic theory. Russ. Math. Surveys 166, 21 - 69 (1972).

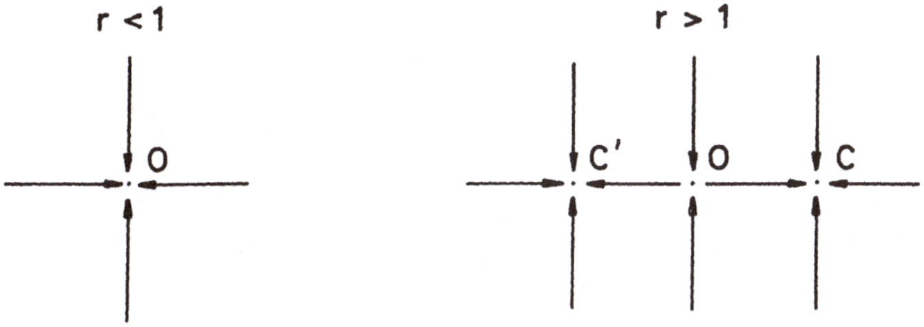

Fig. 1

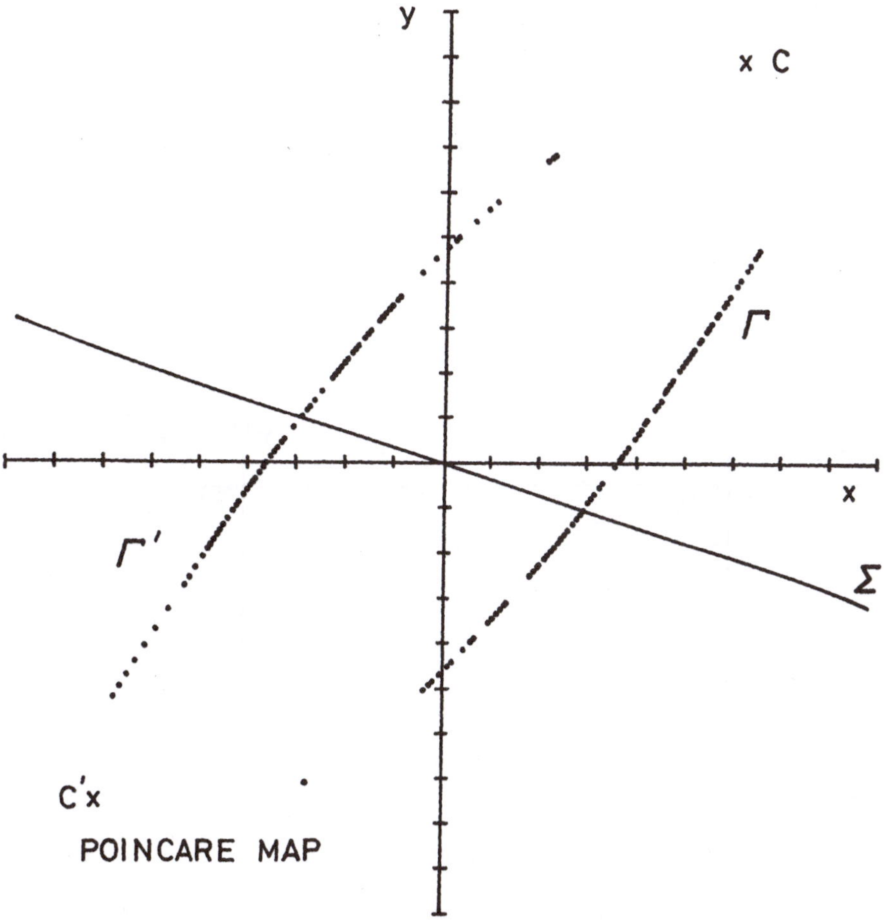

Fig. 2

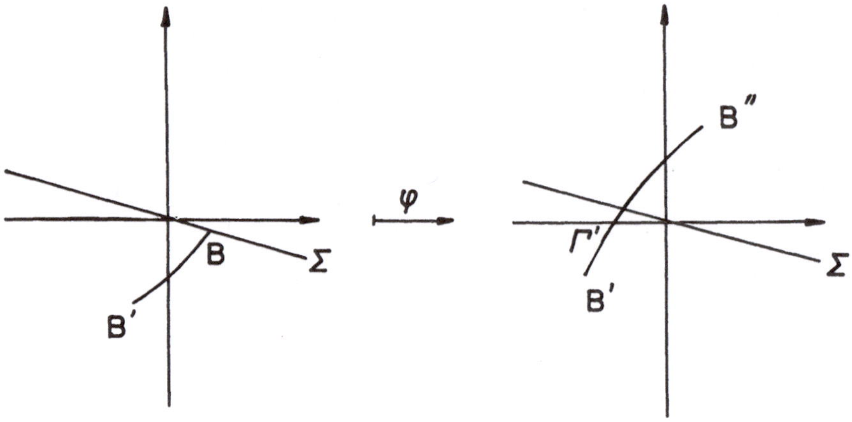

Fig. 4

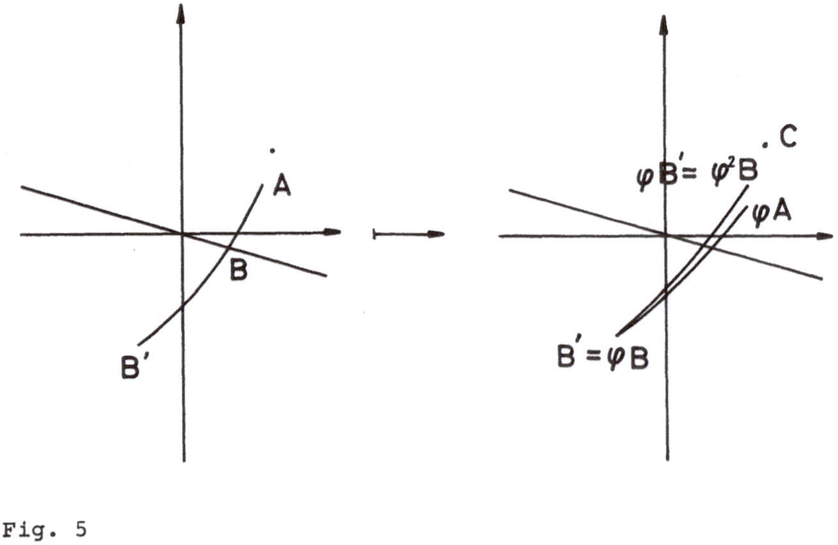

Fig. 5

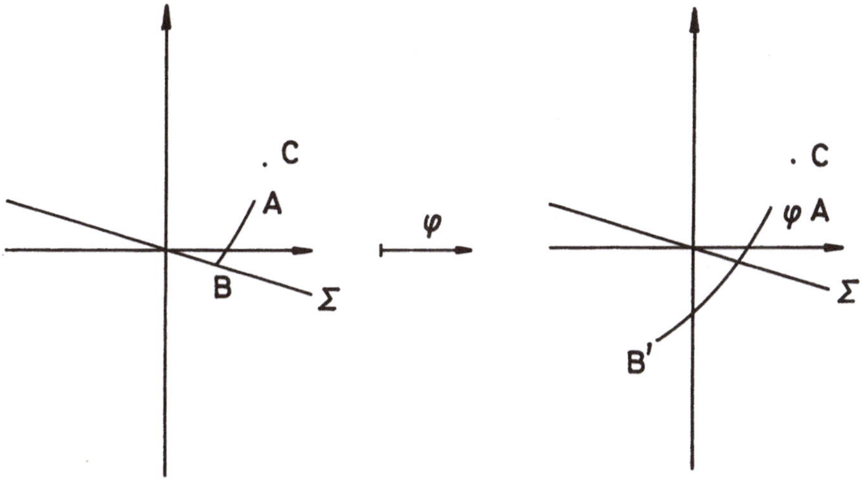

Fig. 3

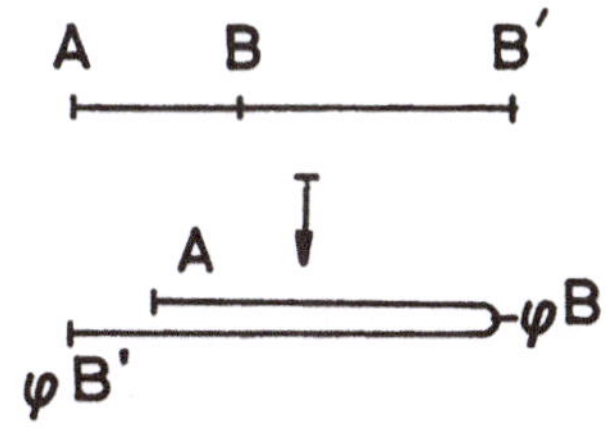

Fig. 6